Biopolymers

Biopolymers
Utilizing Nature's
Advanced Materials

Syed H. Imam, EDITOR
USDA-National Center for Agricultural Utilization Research

Richard V. Greene, EDITOR
USDA-National Center for Agricultural Utilization Research

Baqar R. Zaidi, EDITOR
University of Puerto Rico

Developed from a symposium at the Fifth Chemical
Congress of North America,
Cancun, Quintana Roo, Mexico,
November 11–15, 1997

American Chemical Society, Washington, DC

Library of Congress Cataloging-in-Publication Data

Biopolymers : utilizing nature's advanced materials : developed from a symposium at the Fifth Chemical Congress of North America, Cancun, Quintana Roo, Mexico, November 11–15, 1997 Syed H. Imam, editor, Richard V. Greene, editor, Baqar R. Zaidi, editor.

     p.   cm.—(ACS symposium series, ISSN 0097-6156 ; 723)

    Includes bibliographical references and index.

    ISBN 0-8412-3607-0 (cloth)

    1.   Polysaccharides—Biotechnology—Congresses. 2. Biopolymers—Biotechnology— Congresses. I. Imam, Syed H., 1953- . II. Greene, Richard V., 1954- . III. Zaidi, Baqar R., 1943- . IV. American Chemical Society V. Chemical Congress of North America. (Fifth : 1997 : Cancún, Mexico) VI. Series.

TP248.65.P64B56   1999

660.6′3—dc21

                                                 98-53176
                                                    CIP

CMH

Advisory Board

ACS Symposium Series

Foreword

THE ACS SYMPOSIUM SERIES was first published in 1974 to provide a mechanism for publishing symposia quickly in book form. The purpose of the series is to publish timely, comprehensive books developed from ACS sponsored symposia based on current scientific research. Occasionally, books are developed from symposia sponsored by other organizations when the topic is of keen interest to the chemistry audience.

Before agreeing to publish a book, the proposed table of contents is reviewed for appropriate and comprehensive coverage and for interest to the audience. Some papers may be excluded in order to better focus the book; others may be added to provide comprehensiveness. When appropriate, overview or introductory chapters are added. Drafts of chapters are peer-reviewed prior to final acceptance or rejection, and manuscripts are prepared in camera-ready format.

As a rule, only original research papers and original review papers are included in the volumes. Verbatim reproductions of previously published papers are not accepted.

ACS BOOKS DEPARTMENT

Contents

INDEXES

Preface

OVER THE PAST CENTURY, GLOBAL ECONOMIC ACTIVITIES have increased more than fifty-fold. This extraordinary growth has raised serious concerns about current patterns of production and consumption. Greater attention has been given to the concept of sustainable economic systems that rely on renewable sources of energy and materials. The use of biologically derived polymers (biopolymers) emerges as an important component for economic development. By transforming agricultural feedstocks, or harnessing the enzymes found in nature, a new class of renewable, biodegradable and biocompatible materials is being introduced. Emerging applications for biopolymers range from packaging to industrial chemicals, to medical implant devices, to computer storage media. In addition to producing "green" materials with unique physical and functional properties, the processes used to create biopolymers leads to new manufacturing approaches that minimize energy consumption and waste generation. As the world addresses a growing list of environmental problems, the possibility of using proteins, carbohydrates, and other biopolymers to meet the materials requirements of an expanding economy is likely to receive increasing attention.

A symposium, entitled "Natural Polymers as Advanced Materials," was held at the Fifth Chemical Congress of North America, in ancun, Mexico, November 11–15, 1997, to address technical constraints associated with increased use of biopolymers. That title reflected a degree of naivete by implying that synthetic materials are more advanced than their natural counterparts. After attending the Cancun presentations and after reading the chapters contained in this volume, the Editors are better educated. Hence, the title of this volume was changed to "Biopolymers: Utilizing Nature's Advanced Materials." Nevertheless, from the 33 presentations at the Cancun symposium, came 14 chapters in this volume. Another four chapters were solicited. The chapters, which are a combination of original research and review articles, describe current issues that are dictating biomaterial research directions. We are confident the reader will enjoy their content as much as we did.

We are grateful to the American Chemical Society (ACS) and the ACS Books Department for providing a forum for presenting this work. We particularly thank the authors and the reviewers for their efforts and expertise. We also thank

Proctor & Gamble Company, Rohm & Haas Company, McDonald's Corporation, Biotechnology Research and Development Corporation, Muir Omni Graphics, Monsanto, and ACS for financial support, as well as the Department of Marine Sciences, University of Puerto Rico at Mayaguez and the Agricultural Research Service of the U.S. Department of Agriculture for administrative support. Special thanks go to Deborah K. Bitner for her superior office skills and dedicated professional efforts.

SYED H. IMAM
RICHARD V. GREENE
Agricultural Research Service
U.S. Department of Agriculture
Peoria, IL 61604-3999

BAQAR R. ZAIDI
Department of Marine Sciences
University of Puerto Rico
Mayaguez, PR 00681-9013

OVERVIEW

Chapter 1

Natural Polymers as Advanced Materials: Some Research Needs and Directions

R. L. Shogren[1] and E. B. Bagley[2]

[1]Plant Polymer Research Unit, National Center for Agricultural Utilization Research, Agricultural Research Service, U.S. Department of Agriculture, 1815 N. University Street, Peoria, IL 61604
[2]Biotechnology Research and Development Corporation, 1815 N. University Street, Peoria, IL 61604

At the Fifth Chemical Congress of North America held at Cancun in November, 1997, a broad definition of the fields to be covered in the symposium on natural polymers as advanced materials was given. Included were recent developments and future prospects for natural polymers in uses ranging from detergents, adhesives, packaging, biodegradable plastics, paper, rubber, textile, drug delivery systems to processes including nanometer composites. Several recent articles and books have reviewed the structure, properties and applications of natural polymers (*1-12*). This chapter, rather than reviewing accumulated knowledge of natural polymers, will attempt to point out some shortcomings and opportunities for better understanding and applications. Because of the authors' expertise in starch, many examples of problems and opportunities in this area will be presented as models for similar needs in other areas of natural polymer research.

The concepts involved in polymers as advanced materials hardly come as recent revelations. The advantages of cellulose fibers in paper are not news to technologists. Nature itself has made good use of structural natural polymers, as in chitin, and produced strong polymeric fibers, as in spiderwebs, long before synthetic chemists arrived on the scene to complicate matters. In earlier times, much use was made of natural materials and, especially, carbohydrates. David Morris in an article entitled "Back to the Future" (*13*), maintains that "One hundred and fifty years ago, ours was a carbohydrate economy" with "clothes, houses, vehicles, chemicals, dyes and inks" derived from plants. Products included ethanol, plastics (celluloid and cellophane, the first film plastic) and cotton and wool fibers used in clothing.

---

[3]NOTE: Product names are necessary to report factually on available data; however, the USDA neither guarantees nor warrants the standard of the product and the use of the name by the USDA implies no approval of the product to the exclusion of others that may also be suitable.

Many of the applications of natural polymers lost out in the competitive world to synthetic polymers, in which properties could be tailored to needs. Petroleum-based products had properties independent of the material vagaries associated with crops, whose component properties can vary significantly with growing site, agronomic factors including weather, and other complicated considerations. Molecular weight, degree of branching, and chemical groups in side and main chain sites can be more readily modified and controlled for synthetic polymers than for most natural materials. However, in some cases, chemical modifications of natural polymers can lead to materials of particularly interesting properties; e.g., hydrolyzed starch/polyacrylonitrile graft copolymers with astounding water absorption and holding characteristics...the first superabsorbent (*14*). In general, however, these problems caused natural polymers to become comparatively unattractive at the commercial level. As a result, Morris emphasized that by the 1990's two thirds of our clothes were derived from oil, plastics replaced paper containers, food dyes came from mineral oil and so on. These were all changes indicative of the shift of markets from natural to synthetic materials.

In recent years, the socio-economic situation has changed to make natural polymers once again worth consideration for many applications. First of all, oil embargos and higher oil prices and concern over long-term availability of oil forced technologists and scientists to consider alternative, and especially renewable, sources of materials. Second, environmental considerations made the use of many natural materials very attractive because of their biodegradability, low toxicity and low disposal costs. Finally, properties aside, the use of natural materials, such as starch, for fillers in composite materials was economically attractive because of the high price of synthetic polymer matrices and the low prices of such natural products as starch and starch granules. These low costs are related, of course, to the large agricultural surpluses of recent years and the ability of farmers to produce more. However, utilization of natural polymers only because they are cheap fillers hardly justifies their inclusion as "advanced materials." To classify as advanced materials, there should be significant new and sophisticated technical features in their design, development, or processing. That is, there must be scientific or engineering factors, that go significantly beyond mere economic advantage, to warrant the designation "advanced."

Coverage in Cancun

This symposium addressed many of the current concerns in developing advanced materials from natural polymers. Environmental concerns provided the spark needed to initiate studies of natural materials which would biodegrade readily. One can classify the presentations into work addressing four types of problems: 1)-choice of natural starting material and design of modifications needed for particular applications; 2)-processing to yield products of the needed final properties; 3)-characterization of the biodegradability of the products; 4)-evaluation of mechanisms and reaction pathways.

From the papers presented, it is evident that environmental concerns and particularly issues of product degradability drive much of the research and development work reported at this symposium. Biodegradability can be considered in a broad context as illustrated by the presentation of Maysinger (McGill University) reporting on the use of biodegradable polymers for delivery of neuroactive agents to the nervous system and Heta Ravel's discussion of controlled release of pentachlorophenol from various biodegradable polymer matrix systems. Whether the biodegradability is required for plastics used as mulch on farms or for use in controlled release in the human body or elsewhere, there is a need to know how the materials decay. Hence, considerable attention is being given to the mechanisms of biodegradation. Greene and Imam discussed a working model for predicting the relative rate at which polymer formulations will biodegrade. Other approaches included investigations by Gilkes *et al*. of cellulose hydrolysis by microorganisms.

In terms of materials, *per se*, starch and cellulose must be pretty well put at the head of a list of natural polymers and many of the presentations at Cancun centered on these materials, with starch being studied in both the granular and "thermoplastic" state. Obtaining these materials in fibrous form opens up application opportunities as discussed by Bergthaller *et al*. any of his examples are composites, as are many natural materials such as chitin, wool, etc. Opportunities for utilization of microfibrils and microcrystalline starch and cellulose (a result of biosynthesis of various polysaccharides) in preparation of nanocomposite materials was discussed by Cavaille and Dufresne. For many applications though, the natural polymer must be modified. For control of useful surface active properties, Tao and Wang prepared long-chain fatty acid esters of glucan carbohydrate polymers and oligomers.

Interestingly, lignin, the second most abundant natural polymer in the world (cellulose being number 1), was not covered in a presentation in Cancun. This is probably a reflection of the relatively scant amount of research being done on lignin and the difficulty of working with this complex polymer (*15*). One would expect that lignin and other materials which are now considered "waste" products, such as wheat straw, corn cobs and stalks, soybean hulls, citrus peels and chitin from crustacean shells, will receive more attention in the future as world population and demand for products from natural resources increases.

Proteins of course are a major form of natural polymer and, in combination with synthetic polymers, can provide unique opportunities. Thus, Huang discussed interpenetrating and non-interpenetrating networks of both hydrophilic and hydrophobic protein with synthetic biodegradable polyether and polyester components to produce materials of a wide range of chemical and physical properties. As with other composites containing both natural and synthetic components, the combinations are better than the natural polymer alone. There is considerable current interest too in edible coatings, where proteins can be especially relevant and particular interest was expressed in combinations, such as pectin/starch/calcium.

Problems and Research and Development Opportunities

Molecular and Supramolecular Structures. Surprisingly, there is often a lack of solid information on the molecular properties of natural polymers. Starch, for instance, is a polymer used in enormous quantities in industrial and food applications, but evaluation of molecular parameters is difficult. Amylose is perhaps better characterized than is amylopectin, where quoted molecular weights are into the hundreds of millions (*16,17*). The measurement of molecules showing such large values of molecular weight can be questioned. For example, in light scattering studies, large agglomerations of molecules, either as entangled species (which may even be formed as a result of the polymerization process during biosynthesis) or as a result of actual cross-linking processes resulting in gel particles, can lead to unrealistically high molecular weight values. Attempts to filter the samples for light scattering or in a chromatographic experiment can, no doubt, result in spurious numbers. One can never be sure that what is removed isn't the critical portion of the sample and the issues and concerns are complicated by the fact that processing (as in filtration) could, in principle, result in shear degradation of the molecules. More extensive and more detailed investigations of these points seem well warranted.

One reason for further investigation of molecular and associative properties of these natural polymers is their molecular weights are often so great and they are so "entangled" or agglomerated through physical interactions that processing has a significant effect on properties (*18-20*). The systematic investigation of processing effects requires reliable and sensitive means.

Another area where systematic scientific information is lacking with many natural polymers is in knowledge of the solubilities and phase separation characteristics of natural polymers. The importance of such information has been emphasized for protein systems by Tolstoguzoff and coworkers (*21,22*), but the lack of solid information can also be illustrated again with reference to starch. In Barton's work on polymer-liquid interaction parameters (*23*), this ubiquitous natural polymer is conspicuous by its absence. Amylose receives some minimal attention, but amylopectin is almost completely neglected. The reasons for this are complex. Not least of them is that starch and proteins with polar forces present, along with numerous opportunities for hydrogen bonding, are very difficult to handle theoretically and experimentally. Much of the information available, therefore, is empirical and relatively unsophisticated at that.

There is a great need for extensive fundamental investigations into the molecular structure of natural polymers, their associative and agglomerative behavior, their phase behavior, their solubility characteristics in water and mixed solvents. Much information is needed to provide the background information for materials design, processing and property interpretation. Great opportunities exist for the natural polymer scientist and engineer.

New Natural Polymers and Modeling. Computer based techniques have begun to be used to design and predict the properties of polymers, primarily of the synthetic type (*24*). Some pioneering molecular modeling studies of plant proteins, starch and cellulose have been carried out recently (*25-27*). These have just scratched the surface of what is possible. The potential of emerging computer based methods to predict conformation, interactions and properties of natural polymers, their chemical derivatives and blends is immense. This has become feasible for large molecules only recently as computer speed has increased and molecular dynamics simulations and new approximate methods have been developed.

Due to phenomenal advances in molecular biology, it may well become feasible in the near future to genetically manipulate plants (such as corn or soybeans) to produce natural polymers with well defined molecular weight and branching and even introduce functional groups, such as acetyl, sulphate, amino and phosphate moieties. This may be easiest in the case of proteins, where the amino acid sequence follows directly from the sequence of bases in the plant's DNA. For example, researchers are now modifying genes to improve the mechanical properties of animal proteins, such as elastin, silk and various adhesive polymers (*12*). For other types of natural polymers, the chemical structure is determined from DNA indirectly via the activity of a series of enzymes. Nevertheless, work on expression of bacterial genes for the production of poly(hydroxybutyrates) in plants is progressing and may eventually yield a polyester of very low cost (*28*). New genetic varieties of corn containing starches of different branch frequency and length have recently been developed (*29,30*). The preparation of completely new polymers in plants may even be possible.

This points to the need for more understanding of the relationships between structure and function in natural polymers so we will know what kind of polymer we should make for a given application when the genetic/synthetic machinery is developed. Computer modeling may well become the "screening method" of choice to identify new polymers without having to go through the long and tedious process of synthesizing or "growing" every possibility. Also, it should not be forgotten that we currently raise only a dozen or so plant crops in large amounts and that the hundreds of thousands of other species of plants (not to mention millions of species of microorganisms) represent an undiscovered source of new polymers and fiber. Exciting work for the botanical explorer!

Processing and Composites. Little is known about how the secondary and tertiary structures of natural polymers can be controlled during processing and how processing can be accomplished in a rapid and environmentally friendly manner. Most synthetic polymer based plastics are processed so that molecular orientation and crystallinity are optimal for the intended application. For example, commercially available biaxially oriented polyolefin films have strengths which are several times that of unoriented films. Orientation within the cell walls of polystyrene foam give the product its good flexibility and resilience (*31*). Crystallinity also increases strength, as well as solvent resistance and high temperature stability. Cellulose and proteins can be oriented into fibers, but only by using organic solvents or strong bases to first dissolve the polymers (32). Strong interactions between polar groups make thermal processing difficult for proteins and impossible for cellulose. New plasticizers and/or chemical modifications are needed to improve thermal processibility (i.e., by extrusion). Oriented starch films have never been prepared even though orientation would certainly improve the flexibility and toughness of this normally brittle material. The availability of a totally linear amylose would make orientation easier than for the branched starches now available. Crystallization is rather slow for many biopolymers, such as starch (33) and poly(hydroxybutyrates) (34). Rates of crystallization need to be reduced to seconds to allow economically feasible processing. Hints at ways to control the crystallinity of cellulose may come from current work on use of enzymes to disrupt crystallinity and promote efficient saccharification and fermentation to ethanol.

Several different types of processing techniques have the potential to physically or chemically modify the properties of natural polymers in efficient and environmentally friendly manners. For example, extruders have been used by the food industry and more recently by the biodegradable plastics industry. They rapidly impart thermal and mechanical energy to starch/water mixtures and, thus, transform them into a thermoplastic melt (35). Extruders can be also used as reactors to continuously modify the structure of the polymer (36). Some examples of this include the preparation of hydrolyzed, oxidized, grafted, cationic and succinylated starches (36-39). Potential benefits of reactive extrusion include shorter reaction times, less or no requirement for solvents and fewer byproducts. Other types of processing which utilize microwaves, electromagnetic radiation and ultrasound have yet to be evaluated for most natural polymers.

Most natural materials are actually composites of a fibrous polymer or mineral with an amorphous polymer binder. Examples include wood, insect cuticles, sea shells, and bone. The fiber or mineral filler gives the material high stiffness and strength, particularly since the fibers and minerals are often oriented and arranged in intricate hierarchies. The presence of only a small amount of natural polymer binder in mollusc shell or tooth enamel gives the composite a surprisingly high degree of toughness, when compared to fused mineral alone (40). Reasons for this are not well understood, but may be, in part, due to the ordered arrangement of fibrous mineral or polymer. More work needs to be conducted on interactions which occur between matrix polymers and mineral or fibrous fillers and how these might be controlled to give better adhesion and control of filler morphology (epitaxy).

Significant strides have been made in the processing of synthetic composites, such as glass fiber/epoxy layups, for high performance aerospace and recreational equipment. Progress in the natural polymer area has been slow, although some recent results indicate that this is changing. For example, composites containing >60% granular starch as a filler in poly(hydroxy-ester-ethers) (PHEE) have excellent mechanical properties (41). Unlike other polymers investigated so far, the adhesion between the starch and PHEE is excellent, resulting in granule rupture as the fracture mechanism of the composite. Studies are in progress to determine the molecular basis of the improved adhesion. In other examples, composites of recycled paper and soy flour have a marble-like appearance and have been successfully commercialized. Structural boards based on wheat straw are also being commercialized as a replacement for plywood.

Nevertheless, penetration of new plastics based on natural polymers into the marketplace has been slow because of the high cost of biodegradable polymers which are blended with the natural polymer to improve mechanical properties or the high cost of chemical modifications to the natural polymer (*7,8*). Costs of competing petroleum based polymers, such as polyethylene and polystyrene ($0.40-0.60/lb), are quite low compared to biodegradable polyesters, such as polylactic acid, polyhydroxybutyrate (PHB) and synthetic polyesters ($2-8/lb). Other hurdles also exist, such as the poor water vapor barrier properties of natural polymers due to their polar character. Even PHB, which has the lowest water vapor permeability of the biodegradable polyesters, has a permeability about 10 times that of low density polyethylene (*42*). Yet, natural waxes have some of the lowest water vapor permeabilities known to man. Could the good barrier properties of waxes somehow be combined with the strength of a natural polymer like starch or cellulose? A fatty derivative having a regiospecific substitution pattern, which could arrange its fatty groups in an ordered manner on the surface, might be the answer.

Water Soluble and Swellable Natural Polymers. Most natural polymers are water swellable. Applications that take advantage of this property would seem to be preferred. Since many water soluble polymers are discharged directly into wastewater streams after use, the biodegradability of natural polymers would be an asset in preventing the build up or toxicity in fresh water sources. Water based polymer systems avoid the use of toxic organic solvents, the use of which is becoming more restricted by legislation. Also, the prices of synthetic, water soluble synthetic polymers are quite high, usually over $1/lb. Thus, it is no surprise that starch and cellulose derivatives are widely used as water soluble adhesives, thickeners, binders and coatings (*5,12*). Some emerging applications include polyaspartic acid and oxidized starches as metal ion sequestrants in detergents, mussel proteins as adhesives, cross-linked anionic starches as superabsorbents in diapers and modified agricultural residues as absorbents of organic wastes (*43-48*). More research is needed on basic structure-property relationships of natural polymers to understand their behavior in aqueous systems.

Natural Polymers and Sustainability

In many ways, our current utilization of natural resources cannot be sustained. Most of our fuel for power and transportation comes from fossil fuels, such as oil which will be depleted in the future. Rising atmospheric carbon dioxide levels from combustion of fossil fuels are thought to be increasing global temperatures which, in turn, may cause droughts, crop losses, storm damage, etc. Soil for growing crops is being lost to erosion from wind and rain and the expansion of cities. Irrigation water, particularly in the western U.S., is gradually being depleted as aquifer wells are pumped down. The number of species of living things in the world, estimated at >10 million, are going extinct at a rate of about 0.3% per year (*49*). World population is presently growing at about 1.5% per year, while food production is increasing at only 0.5% per year (*50*). If these trends continue, world population will approximately double to 11 billion in 50 years.

How can natural polymers contribute solutions to these problems? The use of natural polymers as materials and as sources of fuel, such as ethanol, lessens our dependence on non-renewable petroleum. As pointed out by Orts and Glenn in this Symposium, certain natural polymers have the potential of acting as flocculating agents to control soil loss from erosion. Clearly, if the properties of a natural polymer, such as its strength, were improved, less would be needed for a specific application. New applications for agricultural wastes and byproducts of processing, as well as recycling of natural polymers, promise to make more efficient use of our natural resources. Also, alternative crops can produce more biomass per acre of land. Fiber from kenaf and hemp, rather than slow growing trees, can be utilized for a number of applications. This points to the need to conserve plant germplasm since solutions to future problems may lie in species as yet undiscovered.

Literature Cited

1. *Industrial Biotechnological Polymers*; Gebelein, C. G.; Carraher, C. E., Eds.; Technomic: Lancaster, PA, **1995**.
2. *Biodegradable Polymers and Packaging*; Ching, C., Kaplan, D.; Thomas, E., Eds.; Technomic: Lancaster, PA, **1993**.
3. *Biodegradable Polymers and Plastics*; Vert, M.; Feijen, J.; Albertsson, A.; Scott, G.; Chiellini, E., Eds.; Royal Soc. Chem.: Cambridge, MA, **1992**.
4. *Plant Polymeric Carbohydrates*; Meuser, F.; Manners, D. J.; Seibel, W., Eds.; Royal Soc. Chem.: Cambridge, MA, **1993**.
5. *Starch: Chemistry and Technology*; Whistler, R. L.; BeMiller, J. N.; Paschall, E. F., Eds.; Academic Press: Orlando, FL, **1984**, 2nd. ed.
6. *Polymers from Biobased Materials*; Chum, H. L., Ed.; Noyes Data Co.: Park Ridge, NJ, **1991**.
7. Narayan, R. In *Polymeric Materials from Agricultural Feedstocks*; Fishman, M. L.; Friedman, R. B.; Huang, S. J., Eds.; American Chemical Society: Washington, DC, **1994**; pp. 2-28.
8. Mayer, J. M.; Kaplan, D. L. *Trends Polym. Sci.* **1994**, *2*, 227-235.
9. Arthur, J. C. *Polym.Sci. Technol.* **1983**, *17*, 27-40.
10. O'Brien, J. P. *Trends Polym. Sci.* **1993**, *1*, 228-232.
11. Pillai, C. K. S. *Popular Plastics and Packaging* **1996**, *41*, 69-74.
12. *Biopolymers: Making Materials Nature's Way-Background Paper*; U.S. Congress, Office of Technology Assessment, OTA-BP-E-102, U.S. Government Printing Office: Washington, DC, **1993**.

10

13. Morris, D. *The Carbohydrate Economy*, **1997**, *1(1)*, 2-3.
14. Fanta, G. F.; Doane, W. M. In *Modified Starches: Properties and Uses*; Wurzburg, O. B., Ed.; CRC Press: Boca Raton, FL, **1986**; pp. 149-178.
15. Glasser, W. G.; Kelley, S. S. In *Encyclopedia of Polymer Science and Engineering*, 2nd. Ed.; Mark, H. F.; Bikales, N. M.; Overberger, C. G.; Menges, G.; Kroschwitz, J. I., Eds.; J. Wiley & Sons: New York, NY, **1987**, Vol. 8; pp. 795-852.
16. Millard, M. M.; Dintzis, F. R.; Willett, J. L.; Klavons, J. A. *Cereal Chem.* **1997**, *74*, 687-691.
17. Willett, J. L.; Millard, M. M.; Jasberg, B. K. *Polymer* **1997**, *38*, 5983-5989.
18. Lai, L. S.; Kokini, J. L. *Biotechnol. Prog.* **1991**, *7*, 251-266.
19. Sagar, A. D.; Merrill, E. W. *Polymer* **1995**, *36*, 1883-1886.
20. Dintzis, F. R.; Bagley, E. B. *J. Rheology* **1995**, *39*, 1483.
21. Grinberg, V. Y.; Tolstoguzov, V. B. *Food Hydrocol.* **1997**, *11*, 145-158.
22. Polyakov, V. I.; Grinberg, V. Y.; Tolstoguzov, V. B. *Food Hydrocol.* **1997**, *11*, 171-180.
23. Barton, A. F. M. *CRC Handbook of Polymer-Liquid Interaction Parameters and Solubility Parameters*; CRC Press: Boca Raton, FL, **1990**.
24. Kotelyanskii, M. *Trends Polym. Sci.* **1997**, *5*, 193-198.
25. Rothfus, J. A. *J. Agric. Food Chem.* **1996**, *44*, 3143-3152.
26. Trommsdorff, U.; Tomka, I. *Macromol.* **1995**, *28*, 6128-6137.
27. *Computer Modeling of Carbohydrate Molecules*; French, A. D.; Brady, J. W., Eds.; ACS Symposium Series 430; American Chemical Society: Washington, DC, **1990**.
28. Poirier, Y.; Nawrath, C.; Somerville, C. *Bio/Technol.* **1995**, *13*, 142-150.
29. Friedman, R. B. In ref. 1, pp. 13-28.
30. BeMiller, J. N. *Starch/Stärke* **1997**, *49*, 127-131.
31. White, J. L.; Carmak, M. In ref. 15, Vol. 10, pp. 595-618.
32. Rebenfeld, L. In ref. 15, Vol. 6, pp. 647-733.
33. van Soest, J. J. G.; Vliegenthart, J. F. G. *Tibtech.* **1997**, *15*, 208-213.
34. Inoue, Y.; Yoshie, N. *Prog. Polym. Sci.* **1992**, *17*, 571-610.
35. Shogren, R. L.; Fanta, G. F.; Doane, W. M. *Starch/Stärke* **1993**, *45*, 276-280.
36. Carr, M. Proceedings of Corn Utilization Conference IV, National Corn Growers Association: St. Louis, MO, **1992**.
37. Della Valle, G.; Colonna, P.; Tayeb, J.; Vergnes, B. *Spec. Pub. Roy. Soc. Chem.* **1993**, *134*, 240-251.
38. Wang, L.; Shogren, R. L.; Willett, J. L. *Starch/Stärke* **1997**, *49*, 116-120.
39. Wing, R. E.; Willett, J. L. *Ind. Crops Prod.* **1997**, *7*, 45-52.
40. Calvert, P. *Mat. Res. Soc. Symp. Proc.* **1992**, *249*, 539-546.
41. Willett, J. L.; Doane, W. M.; Xu, W. PCT Int. Pat. Appl. WO 9731979, **1997**.
42. Shogren, R. L. *J. Environ. Polym. Degrad.* **1997**, *5*, 91-95.
43. Swift, G. *Polym Degrad. Stabil.* **1994**, *45*, 215-231.

44. *Protein-Based Materials*; McGrath, K.; Kaplan, D., Eds.; Birkhauser: Boston, MA, **1997**.
45. Chan, W. C. *Polym Inter*. **1995**, *38*, 319-323.
46. Laszlo, J. A. *Text. Chem. Color*. **1996**, *28*, 13-17.
47. Wing, R. A. *Starch/Stärke* **1996**, *48*, 275-279.
48. Roweton, S.; Huang, S. J.; Swift, G. *J. Environ. Polym. Degrad*. **1997**, *5*, 175.
49. Wilson, E. O. *The Diversity of Life*; W. W. Norton & Co.: New York, NY, **1992**; pp. 280.
50. Brown, L. R. In *State of the World*; Brown, L. R.; Flavin, C.; French, H., Eds.; W. W. Norton & Co.: New York, NY, **1997**; pp. 23-41.

PROCESSING AND CHARACTERIZATION

Chapter 2

Processing and Characterization of Biodegradable Products Based on Starch and Cellulose Fibers

W. J. Bergthaller[1,4], U. Funke[1], M. G. Lindhauer[1], S. Radosta[2], F. Meister[3], and E. Taeger[3]

[1]Institute for Cereal, Potato, and Starch Technology, Federal Center for Cereal, Potato, and Lipid Research, D-32756 Detmold, Germany
[2]Fraunhofer Institute for Applied Polymer Research, D-14512 Teltow-Seehof, Germany
[3]Thuringian Institute for Textile and Polymer Research, D-07407 Rudolstadt-Schwarza, Germany

Different types of starches, native and chemically modified, were processed in blend systems with natural plasticizers and commercial cellulose fibers by conventional extrusion and injection molding techniques. Processing conditions and quality of products differed widely according to the kind of starch and additives used. Starch polymer characteristics, i.e., amylose/amylopectin ratio and molecular weight distribution, were studied in native and extruded samples in order to reveal effects of structure on product quality. Thus, it was shown that each thermomechanical treatment caused depolymerization of starch, especially of the amylopectin fraction, resulting in decreased tensile strength and loss of product quality, while the amylose fraction remained stable. Considerable improvements in product properties were achieved by adding small amounts of fibers to the starch system. Blending starch with 2-4% fibers resulted in increased tensile strength and water resistance of these products.

In Germany and the European Union technical utilization of starch has a long tradition, particularly in paper and corrugated paper production as the main areas. Other important areas of utilization include chemical uses and fermentation as well (1). For the last twenty-five years advantages and prospects of utilizing starch as a biodegradable component in various combinations with polymers have been discussed with persistence and a lot of energy. The concern over limited capacities in landfill in combination with artificially restricted availability of natural oil as a base material in the production and use of polyolefins in packaging purposes allowed, for the first time, serious considerations of using and testing starch as a natural polymer. Interesting uses

[4]Current address: Federal Center for Cereal, Potato, and Lipid Research, Schuetzenberg 12, D-32756 Detmold, Germany.

14

of starches have been investigated since then, but starch has not yet gained a significant portion of this market.

Nevertheless, starch and cellulose have been studied extensively in order to produce biodegradable materials *(2-4)*. From a chemical point of view and in a certain similarity to cellulose, for which multiple uses in an esterified form are characteristic and well documented, polymeric starch seemed to be an ideal material for reducing the high proportion of biodegradable synthetic packaging materials *(5,6)*.

Destructurization of Starch

Starch granules undergo multiple changes during extrusion processes. Depending on conditions applied, shear forces in particular, the achieved changes may comprise transformations beginning with damage of starch granules and ending with depolymerization of starch molecules. Under extrusion conditions, granular starch can be transformed into a homogenous melt by applying very high heat and shear energy even at low moisture content. During this destructurization process the crystallinity of starch granules is progressively lost *(7,8)*. Simultaneously, mainly mechanically induced depolymerization results in fragmentation of starch molecules that can be described as random chain cleavage and affects amylopectin more than amylose *(7,9)*.

Effects corresponding to depolymerization of crystalline structures happen with native starch granules during transformation into a melt either in blends with various synthetic polymers or as a single component and have been described as destructurization *(10)*. They can be achieved by using single screw extruders or co-rotating twin-screw extruders and applying carefully adapted process conditions as well as by addition of adequate, mostly polar plasticizers. Besides water, the most common plasticizer, various polyfunctional alcohols, such as glycerol, di-, tri-, and polyethylene glycol, sorbitol, sucrose, urea, polyvinyl alcohol etc., were used as additives or additive systems for starch plastification *(3,11-15)*. Recently methyl glycoside *(16)* or selected amino acids in combination with glycerol *(17)* have been studied as new interesting plasticizers, at levels up to 20 or 30%, respectively. As long as compatibility of plasticizers is optimal and their migration minimal a homogenous melt, so-called thermoplastic starch (TPS), is obtained even from granular starch. The degree of destructurization is effected by material based factors, e.g., starch type, total amount of available water, type and portion of plasticizer on the one hand, and by process variables, e.g., screw speed and configuration, and barrel temperature profile on the other. A general representation of potential parameter effects of destructurization of starch and formation of thermoplastic starch is given in Figure 1. The resulting polymeric material is hydrophilic, flexible, biodegradable and can be processed thermoplastically into films and molded pieces by using conventional synthetic polymer processing machines *(15,18)*.

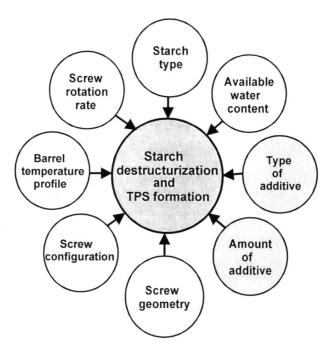

Figure 1. Parameter effects of destructurization of starch and formation of thermoplastic starch - TPS (Adapted from ref. 15).

TPS as Composite Matrix

TPS products have been described as promising and competitive materials to common thermoplastic compounds. Their tensile strength values of 18 to 25 N•mm^{-2} at equilibrium moisture conditions and plasticizer contents of 15-20% glycerol and also their elastic moduli of 1200-2400 N•mm^{-2} exceeded even those of polyethylene. An extended study *(15)* showed that tensile characteristics can be improved further by incorporating short flax fibers (fiber length: 20-50 mm) to TPS. By applying 20 to 30% of flax fibers tensile strength values of 24 to 44 N•mm^{-2} have been achieved in starch/water and starch/glycerol systems. However, elongation characteristics were reduced by fiber addition. In any case, an increasing amount of added glycerol affected inversely the degree of destructurization *(15, 18)* and as a consequence reduced the binding capacities of embedded fibers resulting in a decrease in mechanical properties.

The aim of the present study was to initially investigate structural changes of different native and hydroxypropylated starches depending on extrusion conditions, i.e., effects of mass flow, in particular starch types and plasticizer addition, amounts of water in the starch/water system, screw configurations, and screw rotation rates. The starch samples were characterized by their amylose content and their molecular weight as well. The input of specific mechanical energy (SME) was taken as a measure for the formation of homogeneously destructurized starch (TPS). Decomposition of starch granules was observed by scanning electron microscopy. Besides, the degradation of starch molecules was investigated by gel permeation chromatography in order to point out fragmentation processes occurring during extrusion.

Subsequently, the effect of addition of two types of purified, short natural cellulose fibers was studied in compounding and injection molding of different native and hydroxypropylated starch types. The structural similarity of cellulose and starch polymers should allow the formation of H-bridges between both molecular types, consequently improving adhesion. During formation of fiber-reinforced TPS the above mentioned system and target parameters were used again. In order to validate the practicality of the starch/fiber composites mechanical (tensile strength, elongation) and chemical (water adsorption and water solubility) property profiles were taken from injection molded test pieces.

Materials and Methods

Starch Types. From the scale of available commercial starches of different origin a series of native and hydroxypropylated starches were selected for destructurization experiments and for preparation of polysaccharide blends. Their equilibrium moisture content varied between 10 and 15% at standard conditions and was taken into account while calculating formulations. They have been characterized by amylose content and molecular weight (Table I).

Additives. In TPS-formation experiments different types of commercially available low molecular polyalcohols, such as sorbitol (Sorbidex, Cerestar Deutschland GmbH, Krefeld, Germany) or glycerol (Weinstock & Sievert, Düsseldorf, Germany), have been used as plasticizers.

Fiber Material. In compounding experiments two samples of purified, short natural cellulose fibers (a ligneous type originating from south pine tree and a cotton type) were applied. Table II contains chemical and physical characteristics of both fiber types.

Table I. Selected Starch Types for Production of Polysaccharide Blends

Source Material	Brand Name	Amylose Content %	Molecular Weight 10^6 g/mol
Native Types			
Waxy maize[a]		0	40
Potato[b]		23	40-50
Maize[a]		20	40-50
Amylomaize[c]		71	6.0
Modified Types			
(Hydroxypropylated, DS 0.05-0.08)			
Potato[b]	Emzise E2	19	9.9
Amylomaize[d]	Gelose 22	36	9.2
High-amylose maize[d]	A 965	62	3.8

[a]Cerestar Deutschland GmbH, Krefeld, Germany.
[b]Emsland-Stärke GmbH, Emlichheim, Germany.
[c]National Starch and Chemical, Weinheim, Germany.
[d]Starch Australasia, Ltd., Lane Cove, NSW, Australia.

Extrusion. For destructurization of starches a closely intermeshing co-rotating twin-screw extruder (type ZSK 40, Werner & Pfleiderer, Stuttgart, Germany) having 40 mm diameter screws and a 28/1 L/D ratio was used. Adapted screw configurations of modular construction and barrel temperature profiles within six heating zones (T_1 to T_6)

are documented in Figure 2a. The feeding zone consisted of two barrel segments. The processing zone for destructurization and blending of components comprised five barrel segments. Barrel segment temperatures were followed by means of NiCrNi thermo-couples. The temperature of destructurized starch samples was measured in the gap in front of the die. The pressure of the destructurized starch was measured there, too, by means of a sealed force transducer (type MDA 420, Dynisco Geräte GmbH, Heilbronn, Germany). Further experimental conditions are documented in Table III.

Table II. Fiber Characteristics of Samples Used in Starch Compounding

Brand Name/ Type	Purity %	Length mm	Thickness μm	DP	Tensile Strength at Break (cN/tex)
Cellunier F (ligneous fibers)	94-97	4 (3.0-5.5)	40 (35-45)	564	36-40
Temming 500 (cotton fibers)	97-99	1 (0.5-1.7)	20 (8-32)	1634	30-32

Table III. Experimental Conditions in Extrusion Processing

Extrusion Parameters		Destructurization	TPS-Fiber-Blend Formation
Screw Rotation rate • minutes^{-1}		90 - 240	120
Die temperature	°C	74 - 98	82 - 94
Mass pressure	bar	5 - 48	16 - 26

In extrusion experiments starches were fed via a volumetrically operating twin screw feeder providing a mass flow of 30 kg•h^{-1}. Water and glycerol were administered from separate supply tanks by means of two piston metering pumps and fed at the second barrel section as a combined stream. Samples of approx. 1 to 2 kg of destructurized starch were collected after adjustment of process parameters and a minimum operation time of 5 min at constant conditions. Relevant data of extrusion were registered automatically within an interval of 30's during each sample production. A representative set was printed then as a protocol. Values of specific mechanical energy (SME) input were calculated by using equation 1 on the basis of the electrical power input, the relative effective screw rotation rate, the relative effective torque, and the total mass flow (of solid and liquid components) *(26)*.

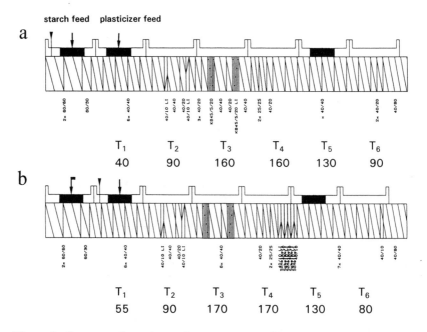

Figure 2. Screw configuration and temperature profiles (T_1 to T_6) for extrusion experiments (a, destructurization; b, TPS-fiber-blend formation).

$$SME = \frac{P_{max}}{\dot{m}} \cdot \frac{T_{eff}}{T_{max}} \cdot \frac{N_{eff}}{N_{max}} \cdot 3600 \qquad (1)$$

SME Specific mechanical energy [kJ·kg⁻¹]

P_{max} Maximum motor power [Wh] (34,5 kW)

$\dot{m}$ Mass flow [kg·h⁻¹]

T_{eff} Effective torque of screws [%]

T_{max} Maximum torque of screws [%]

N_{eff} Effective screw rotation rate [min⁻¹]

N_{max} Maximum screw rotation rate [min⁻¹] (400 min⁻¹)

Compounding of fiber reinforced starch blends was done using the extrusion equipment described above and the screw geometry shown in Figure 2b. Relative portions of components in TPS-fiber-blend formulations were as follows: Starch and fibers represented 67% of the total mixture. Sorbitol and glycerol were 13% and 20%, respectively. Gelose 22, a hydroxypropylated maize starch (Table I), was used as basic material. Although Fritz et al. *(15)* described the incorporation of 20 to 30% of short flax fibers the two types of cellulose provided in this study (Table II) did not allow addition of such high amounts because of their smaller bulk density. They were varied, therefore, in portions of 2, 4, 7, and 15%. Since formulation should be held constant the solid components (starch, sorbitol and fibers) were carefully premixed in a rotating drum for 15 min. With increasing amount of fiber materials the mass flow of the carefully mixed solid materials (starch, fiber, and sorbitol) was reduced to 10.5 kg•h⁻¹ and the mass flow of the liquid components (water, glycerol) to 3.9 kg•h⁻¹. The total mass flow had to be reduced to a minimum of 14.4 kg•h⁻¹.

Injection Molding. For thermoplastic processing an injection molding machine, Type 170 CMD (Arburg, Loßburg, Germany), was used. The injection unit consisted of a barrel divided into 3 heating segments and a separately heated die segment. It was equipped with a 25 mm diameter single screw (L/D ratio = 15) having constant pitch and unchangeable geometry. The temperature profile along the barrel including the die provided four temperature zones in rising order beginning immediately after the feeding section with temperatures of 160, 170, 180, and 190°C. A heated mold was used for production of standard test pieces of 150 mm length (DIN 53455) *(19)*. The injection molding cycle was set for 30 to 35 seconds in total from which 0.9 to 1 second was needed for injection, 3 to 5 seconds for pressing, 15 seconds for cooling, and 7 to 13 seconds for dosage.

Weight Average Molecular Weight Determination. The weight average molecular weight of starch samples and extrudates was determined by the HP-SEC-MALLS method after their solubilization in DMSO at a concentration of 0.1%. Starch and DMSO were weighed in a Schott bottle so that the starch concentration was 5 mg•ml^{-1}. The solution was stirred at 80°C for 8 hr and then at 100°C for 2 hr with a magnetic stirrer (300 min^{-1}). Nitrogen was applied to prevent oxidative degradation. 100 µl of the solution were taken for the analysis. The HP-SEC System (Waters) consisted of a 600MS pump module, 717 autoinjector, column compartment, RI-Detector 410, and MALLS-detector. The MALLS detector was a dawn-F-DSP laser photometer (Wyatt Technology, Inc., Santa Barbara, USA), fitted with a S2 flow cell and a Ar-ion laser operating at λ= 488 nm and equipped with 18 detectors at angles ranging from 7.5 to 157°. The columns were Waters Styragel HMW 7 (effective molecular weight range 10^8 to $5•10^5$), HMW 6E (10^7 to $5•10^3$), and HT 3 ($3•10^4$ to $5•10^2$) with a dimension of 300 mm•7.8 mm. The elution of samples was carried out with DMSO with 0.09 m NaNO$_3$ at a flow rate of 0.5 ml•min^{-1} and a temperature of 60°C.

The MALLS-detector was serially connected with the refractive index detector (DRI). During a sample run on the HP-SEC-MALLS system the data from the DRI and MALLS detector were collected and processed to give the molecular weight at each retention volume.

The correlation between scattered light from a diluted macromolecular solution and the molecular weight distribution was formulated by Debye *(20)* and Zimm *(21)*. On the basis of this relation Wyatt *(22)* described the absolute characterization of macromolecules by HP-SEC-light scattering experiments. Molecular weight and radius of gyration were calculated for each slice of the chromatogram on the basis of usual light scattering equations using the ASTRA 4.2 software.

Scanning Electron Microscopy. The fine structure of cross-sectional surfaces of extruded cords and injection-molded test specimen was investigated using a scanning electron microscope (type Leitz AMR 1600 T, Leitz KG, Wetzlar, Germany). Sample preparation was carried out after slow freezing with liquid nitrogen by carefully breaking the frozen specimen. Characteristic details have been documented photographically.

Mechanical Properties. Based on a standardized procedure of a tensile test, DIN 53 455 *(19)*, tensile strength and elongation of extruded cords and injection-molded test specimen were determined by using a universal testing machine (type 14260, Zwick GmbH & Co., Ulm, Germany). From each sample 10 test pieces of 150 mm length were used for measurement. Prior to measurement all test pieces were conditioned by storing over a period of 14 days at 25°C and 45% relative humidity. The test speed was set to 100 mm•s^{-1}. Maximum tensile strength and percent elongation at break were derived from stress-strain diagrams and used for calculation of corresponding average values.

Water Adsorption. Hydrophilicity of products was determined by storing the granulated extrudates after drying for 30 min at 120°C in 45% relative humidity and analyzing the time dependent weight increase over a period of 14 days *(23)*. The equilibrium moisture content was generally reached within a period of 12 days and was used for calculation of the percent water absorption.

Water Solubility. Solubility was measured by using a standard procedure *(23)*. Approximately 0.5 g of dried granulated extrudates were suspended in 50 ml cold tap water and stirred for 20 hr. Then the insoluble part was separated by filtration, dried and weighed and used for calculation of the percentage of soluble material.

Results and Discussion

Destructurization of Starch Types.

Effect of Water Addition on SME. As described previously water is a well known plasticizer in the extrusion of starch. Its variation clearly affects certain system parameters, in particular the temperature of the destructurized starch and the pressure at the die head.

In experiments described in the following the mass flow was set for 30 kg•h^{-1} and the screw rotation rate to 120 per min. The range of water addition began with approximately 11% (mass flow: 3.6 kg•h^{-1}) and ended with approximately 20% (mass flow: 7.2 kg•h^{-1}). With all starch types the increase of added water resulted in a steady decrease of torque and in consequence a decrease of calculated specific mechanical energy (SME) (Figure 3). The observed effect of an approximately linear decrease was most pronounced with the amylopectin rich potato starch, while more amylose rich starch types gave less high SME levels and smaller reduction rates. The modification of starch with hydroxypropyl groups established in all cases a lower level of energy input, but some of the observed differences vanished with increasing amounts of water in the system.

With native as well as modified potato starch (Emsize E2) a high level of mechanical energy input was reached even at a rather high amount of water addition (SME >300 kJ•kg^{-1}). Reaching this level allowed the formation of a homogenous system with fully destructurized starch. In case of maize, amylomaize starch, and the hydroxypropylated amylomaize starch (Gelose 22) SME values fell below this level at water additions of 14 to 17% depending on starch type. For the hydroxypropylated high amylose maize starch (A 965) an amount of water far below the tested minimum water addition (11%) seemed to be necessary for generation of sufficiently destructurized starch (Figure 3).

Effect of Screw Rotation Rate on SME. At a given mass flow (30 kg•h^{-1}) including constant water flow (6 kg•h^{-1}) a further possibility of affecting SME directly is given by changing the screw rotation rate. Beginning with 90 rotations per min its rate was increased to 240 rotations without overloading the engine. The increase in

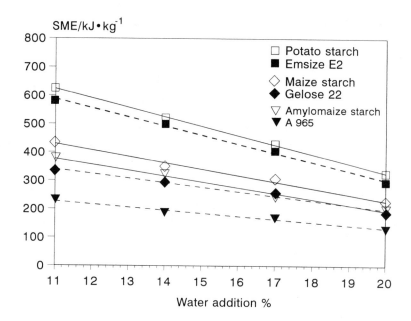

Figure 3. Effect of water addition on specific mechanical energy (SME) input as a function of starch type.

screw speed was accompanied by increasing energy dissipation and rising material shear, which resulted in a significant temperature rise in destructurized starch in front of the die. The torque, however, remained nearly unchanged. The differing starch types showed slightly different curves of SME values (Figure 4). A characteristic nearly linear increase of SME values could be observed with rising rotation rates. The highest SME values of approx. 580 and 500 kJ•kg^{-1}, resp. again were produced with native and hydroxypropylated potato starch (Emsize E2). The smallest values, in contrast, were achieved with native and mod'fied high amylomaize starch (A 965). The pronounced differences in SME levels of t..e tested starches were expected as a result of their amylose/amylopectin ratio and the differing viscosity behavior of the starches used (27). The differences observed between native and hydroxypropylated starches could also be explained by viscosity differences between native and modified starch types and furthermore by defect granular structures in the case of substituted starch types. Nevertheless, as proven in microphotographs of extrudate surface structures (Figure 5b) the extent of destructurization necessary for formation of a homogenous melt could be finally achieved with all tested starches within the experimental range of screw rotation rates.

Molecular Dimensions of Starch Destructurization. The effects of extrusion procedures on the molecular weight of a series of native and hydroxypropylated starches and extrudates thereof have been summarized in Table IV. Potato and maize starch had weight average molecular weights between 40 and 50•10^6 g•mol^{-1}. The reduction of the weight average molecular weight of the samples as a measure of depolymerization depended on the screw rotation rate in extrusion processing, in particular on respective values of the applied specific mechanical energy (SME) input. The higher the SME input the more pronounced was the degradation of starch. In comparison to potato and amylomaize starch the waxy maize starch sample showed the most striking effect on the weight average molecular weight (Table IV).

A comparison of molecular weight distribution curves of different starch types showed, furthermore, that extrusion of native starches resulted preferentially in a degradation of the amylopectin fraction of the starches. This was demonstrated for example in Figure 6 for amylomaize starch. The peak in the high molecular weight region represents the amylopectin component. This peak was drastically reduced whereas the amylose peak remained nearly unchanged.

Extrusion of a chemically modified starches like hydroxypropylated amylomaize starch did not show the same severe degradation effect on the molecular weight of the amylopectin in comparison to the corresponding native starches. Nevertheless a considerable degradation of the amylopectin component was observed in chemically modified amylomaize starch, but the molecular weight fraction of amylose remained nearly unchanged.

Surface Structure of Extruded Starch Observed by Scanning Electron Microscopy. The most important problem of TPS production consisted in forming a

26

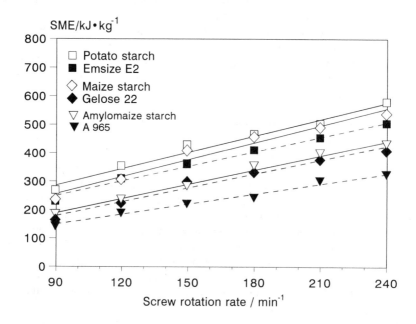

Figure 4. Effect of screw rotation rate on specific mechanical energy (SME) input as a function of starch type.

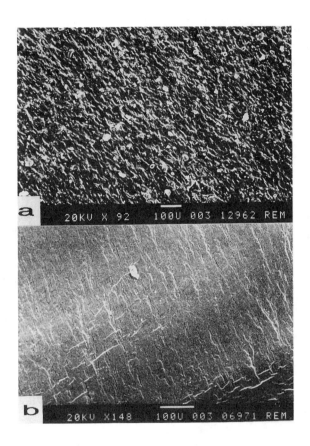

Figure 5. Surface structure of extruded starch (hydroxypropylated maize starch). a, SME = 188 kJ•kg^{-1} (partially destructurized); b, SME = 287 kJ•kg^{-1} (fully destructurized).

Differential weight fraction

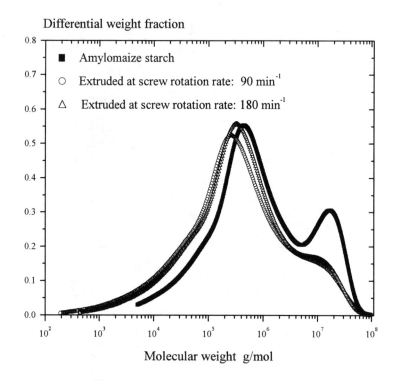

Figure 6. Molecular weight distribution curves of native and extruded amylomaize starch.

homogenous melt of different granular starch types *(15)*. Total transformation of starch granules into a melt becomes visible by a characteristic smooth surface. The degree of destructurization of starch granules was visualized in microphotographs of broken surfaces of extruded TPS cords. The surface structures of extrudates produced under varied conditions differed predominantly with the input of specific mechanical energy. In comparison to an insufficiently managed extrusion process (SME = 188 kJ•kg^{-1}) with hydroxypropylated maize starch for which a rough surface structure was formed by partially destructurized starch granules (Figure 5a) the corresponding surface of a properly extruded sample of the same starch looked rather smooth. In this case a specific mechanical energy input of nearly 300 kJ•kg^{-1} has been applied for total destructurization (Figure 5b).

Table IV. Weight Average Molecular Weights of Starches and Extrudates

Starch Type	Screw Rotation Rate [minutes^{-1}]	SME [kJ•kg^{-1}]	Weight Average Molecular Weight (M_w) [10^6 g•mol^{-1}]
Potato, native			ca. 40 to 50
extruded	90	213	16.7
extruded	180	461	15.2
Maize, native			ca. 40 to 50
extruded	90	179	22.7
extruded	180	459	13.4
Amylomaize A943, native			6.0
extruded	90	190	2.7
extruded	180	345	2.5
Hydroxypropylated amylomaize A965			3.8
extruded	90	148	2.6
extruded	180	243	2.4
Waxy maize, native			40.0
extruded	180	670	5.0

Fiber Reinforced TPS

Effects of Fiber Addition on SME. Based on the previously described extrusion procedure and the compound formulation consisting of 67% Gelose 22, 13% sorbitol, and 20% glycerol portions of 2 to 15% of the modified starch were replaced by the cellulose types described in Table II. Since glycerol content was related to the portion of starch its addition was reduced likewise with increasing cellulose portions. Although in this way a significant reduction of the mass flow occurred, a steady increase of torque was measured which resulted, with one exception, in remarkable rises of corresponding SME values (Table V). The corresponding energy increase comprised a range of 80 to 250 kJ•kg^{-1}. Additional energy increases (100 to 200 kJ•kg^{l}) were achieved by changing screw geometry. In particular, some feeding elements were replaced by reversed-pitch elements. The observed total energy increases proved to be prerequisite for complete destructurization in case of fiber reinforced blends (Figure 7b).

Table V. Results of Extrusion Experiments in the Production of Starch Blends with Different Cellulose Fiber Types

Fiber Type	Mass Flow kg•h^{-1}				Torque %	SME kJ•kg^{-1}	Extrudate Formulation[a] %
	Solids	Glycerol	Water	Total			
Reference	21.0	6.6	0.6	28.2	25	287	67/13/ 0/20
Cellunier F	21.0	6.6	0.6	28.2	33	379	65/13/ 2/20
Cellunier F	15.3	4.8	0.6	20.7	29	454	60/13/ 7/20
Cellunier F	10.5	3.3	0.6	14.4	15	337	52/13/15/20
Temming 500	21.0	6.6	0.6	28.2	32	368	65/13/ 2/20
Temming 500	19.2	6.0	0.6	25.8	36	452	60/13/ 7/20
Temming 500	16.2	5.1	0.6	21.9	36	533	52/13/15/20

[a]Extrudate Formulation: Gelose 22/Sorbitol/Fiber/Glycerol.

Structure of Fiber Reinforced TPS Observed by Scanning Electron Microscopy. In order to visualize embedding capacities of different fiber materials in the starch matrix SEM photographs were used for analysis. In cases where insufficient starch destructurization occurred during extrusion, processing fibers were only slightly embedded in the starch matrix. As a result enhancement of mechanical properties could not be expected. The upper microphotograph of Figure 7a shows the typical rough surface structure of corresponding starch/fiber compositions. Reinforcement, however, was achieved by varying extrusion conditions in order to raise the specific mechanical energy input as described above. A minimum of SME input of nearly 400 kJ•kg^{-1} was necessary for fiber incorporation and total destructurization of starch granules in the case of hydroxypropylated amylomaize starch (Gelose 22). Irrespective of fiber

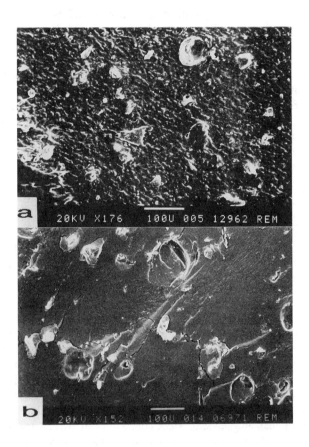

Figure 7. Surface structure of fiber reinforced TPS (Gelose 22). a, SME = 227 kJ•kg^{-1} (partially destructurized); b, SME = 368 kJ•kg^{-1} (fully destructurized).

addition used for reinforcement this in fact became visible by the characteristic smooth surface layer of a corresponding transverse section of an extrudate cord (Figure 7b).

Injection Molding of Standard Test Specimen. Fiber reinforced TPS was cut into granulates immediately after extrusion by rotating blades or after cooling. The dried free flowing granulates of 2 to 3 mm diameter were used for further processing by injection molding. At optimized processing conditions, i.e., plastification temperatures of 170 to 200°C and injection pressures of 1300 to 1500 bar, injection molding resulted in a continuous processing of test pieces within 30 to 35 seconds for each cycle. Increasing amounts of fibers also increased the dosage time of each cycle irrespective of the fiber type used. In comparison to conventional thermoplastics the processing period observed for fiber reinforced TPS lasted several times longer. This was the result of the relative high cooling temperature (70°C) of the molded pieces which could not be reduced further.

Also, shrinking effects, another important and undesired characteristic in processing thermoplastics, were measured in injection molded TPS. Depending on processing conditions shrinking levels reached 2 to 3.3% in TPS specimens without fiber addition. In polyolefins shrinking can achieve values up to 4% *(24)*. In TPS the addition of 2% fiber reduced shrinking to 1.3% or less. Further addition of fiber prevented any shrinking in injection molding. Maximum form stability in starch/fiber compounds was observed in test specimen with 7% fiber addition.

Stress-Strain Behavior. Measurements of test specimens produced by injection molding from TPS based on Gelose 22 gave tensile strength values of approximately 3.5 $N \cdot mm^{-2}$ and percent elongations of approximately 80%. As a prerequisite of the formation of a homogenous TPS matrix a sufficiently high level of SME had to be built up, in particular approx. 300 $kJ \cdot kg^{-1}$. Incorporation of cellulose fibers resulted in an increase in tensile strength with increasing amount of added fibers, but at the same time elongation was remarkably reduced. These effects differed to a certain extent according to the fiber type used (Figure 8). In reinforcement by the ligneous fiber, Cellunier F, a maximum tensile strength of approx. 6.5 $N \cdot mm^{-2}$ was reached with 7% fiber addition. At the same time, elongation at break was reduced drastically to 47%. A further reduction of strain was observed in test specimens with 15% fiber content. In the case of Temming 500, a much smaller cotton type fiber, the enhancement in mechanical strength became more obvious. At 15% fiber addition tensile properties could be increased further to approximately 7 $N \cdot mm^{-2}$. The reduction of product elongation was less pronounced in Temming reinforced test specimens (2% and 7%, resp.), but reached finally the same extent as Cellunier test pieces in the case of 15% fiber addition.

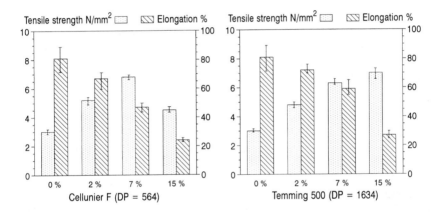

Figure 8. Mechanical properties (tensile strength at break, elongation) of fiber reinforced thermoplastic starch blends. Cellunier F: ligneous fiber type; Temming 500: cotton fiber type (Reproduced with permission from ref. 25. Copyright 1998 Elsevier Science, Ltd.).

The observed differences in fiber reinforcement can be explained by the size of both fiber types and their distribution within the matrix at a given concentration. Reaching a concentration of 15% the bigger particles of Cellunier F might function rather as faults than as reinforcing elements. Besides, the smaller specific fiber surface of Cellunier F is expected to reduce adhesion.

In agreement with results of Fritz et al. *(15)* the incorporation of the applied cellulose fibers improved the tensile strength of TPS considerably. Compared to literature, wherein usually 20 to 30% fibers are used, the experimental results showed that a remarkable enhancement of mechanical strength (doubling of tensile properties) can be achieved even at low fiber content (7%) after adoption of optimized processing conditions. However, the absolute level of the presented results was approximately 8 to 5 times smaller as described there. The observed difference can be easily explained by the much higher plasticizer content (sorbitol/glycerol) used in the experiments presented above. Other potential factors might be found in the differing fiber and starch types. Gelose 22, a hydroxypropylated high amylose maize starch has been used here. In contrast, Fritz et al. *(15)* selected commercial maize starch. The reduction in the elongation characteristic (elongation at break) observed in both studies reached a comparable percent level.

Water Adsorption and Water Solubility. The sensitivity of granulated starch extrudates to water was slightly reduced as soon as these extrudates were reinforced with rather small amounts of cellulose (2 to 7%). However, this effect reversed with further addition significantly beyond concentrations mentioned previously, for instance 15% (Figure 9). The 15% portion of ligneous cellulose (Cellunier F) produced a water sensitivity exceeding respective characteristics of TPS. This pattern could be found in measurements of cold water solubility, likewise, where samples were soaked in water during a period of 20 hr (Figure 10).

Conclusions

In applying carefully adapted extrusion conditions and using water or polyfunctional polyalcohols as natural plasticizers granular starches of normal and high amylose types as well as corresponding hydroxypropylated variants were transformed into thermoplastic starch (TPS). In a co-rotating intermeshing twin-screw extruder the concomitant process of destructurization is dominated by the energy input, in particular the specific mechanical energy (SME) input. Irrespective of the starch type tested the energy input necessary for this kind of transformation could be effected significantly by linear reduction of available water and a likewise linear increase in screw rotation rate. Concerning the screw rotation rate, however, the starch types showed significant differences along a falling line from potato starch to high amylose corn starch and finally ending with slightly hydroxypropylated (DS: 0.08-0.05) high amylose maize

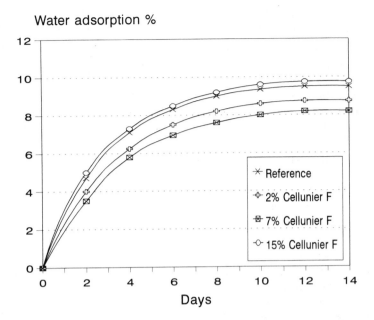

Figure 9. Effect of fiber content on time dependent water absorption of thermoplastic starch blends (ligneous fiber type) (Adapted from ref. 25).

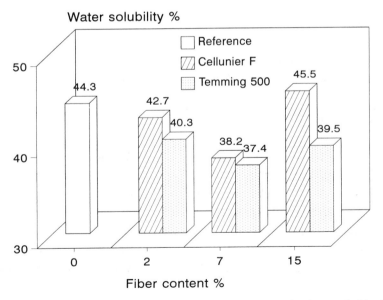

Figure 10. Water solubility of fiber reinforced thermoplastic starch blends depending on fiber content and fiber type.

starch. Correspondingly, the increase of SME was less pronounced with high amylose starches and proved to be reduced even further in cases of hydroxypropylated types.

With regard to molecular dimensions of destructurized starch weight average molecular weight determinations by the HP-SEC-MALLS system as well as molecular weight distribution curves derived from HP-SEC, experiments showed that destructurization produced significant shifts to low molecular fractions. Predominantly the peaks of amylopectin components were reduced whereas amylose fractions remained nearly unchanged. The described changes were well correlated with screw rotation rate or SME, respectively.

In blending TPS produced from hydroxypropylated high amylose corn starch with short natural cellulose fibers (ligneous type: Cellunier F; cotton type: Temming 500) its portion increased SME. Nevertheless, complete destructurization could be obtained within a SME range of 350 to 600 kJ•kg^{-1} when 2 to 15% of fiber material was added.

Compounding of TPS with flax fibers has been described as a method for improving mechanical properties (15) and could be realized here with much smaller amounts of natural fibers. Tensile strength was doubled from approximately 3 to 7 N•mm^{-2} by fiber addition of 7%. Elongation, however, declined extremely, i.e., from 80 to 25%. Besides, additional positive effects could be found in water resistance. Water absorption and water solubility were reduced even at low fiber content. From approximately 7% upwards this effect turned negative, with some differences between tested fiber types. The maximum moisture uptake was 8% within 10 days of storage. Water solubility was reduced by 15 to 20%.

Acknowledgment

This research was made possible through financial support by the Fachagentur Nachwachsende Rohstoffe e.V. (FNR), Güstrow, Germany.

Literature Cited

1.   Anonymus, Stärke - Fortschritt durch Tradition 1946-1996; Fachverband der Stärkeindustrie: Bonn, Germany, **1996**.

2.   Woelk, H. U. *Starch/Stärke* **1981**, *33*, 397-408.

3.   Röper, H.; Koch, H. *Starch/Stärke* **1990**, *42*, 123-130.

4.   Doane, W. M. *Starch/Stärke* **1992**, *44*, 293-295.

5.   Fritz, H.-G.; Seidenstücker, T.; Bölz, U.; Juza, M.; Schroeter, J.; Endres, H.-J. Study on Production of Thermoplastics and Fibres Based Mainly on Biological

Materials. European Commission Directorate-General XII, Science, Research and Development: Brussels **1994** (EUR 16102 EN).

6.   Korn, M. Nachwachsende und Bioabbaubare Materialien im Verpackungsbereich. Roman Kovar Verlag: München, **1993**.

7.   Colonna, P.; Tayeb, J.; Mercier, C. In *Extrusion Cooking*; Mercier, C.; Linko, P.; Harper, J., Eds.; American Association of Cereal Chemists, Inc.: St. Paul, MN, **1989**; pp. 247-319.

8.   Harper, J. M. In *Developments in Carbohydrate Chemistry*; Alexander, R. J.; Zobel, H. F., Eds.; American Association of Cereal Chemists, Inc.: St. Paul, MN, **1992**; pp. 37-64.

9.   Lillford, P. J.; Morrison, A. In *Starch: Structure and Functionality*; Frazier, P.J.; Richmond, P.; Donald, A. M., Eds.; The Royal Society of Chemistry: Cambridge, **1997**; pp. 1-8.

10. Della Valle, G.; Kozlowski, A.; Colonna, P.; Tayeb, J. *Lebensm. Wiss. u. Technol.* **1989**, *22*, 279-286.

11. Stepto, R. F.; Tomka, I. *Chimia* **1987**, *41(3)*, 76-81.

12. Stepto, R. F.; Dobler, B. EP 0.326.517, **1989**.

13. Shogren, R. L.; Swanson, C. L.; Thompson, A. R. *Starch/Stärke* **1992**, *44*, 335-338.

14. Potente, H.; Rücker, A.; Natrop, B. *Starch/Stärke* **1994**, *46*, 52-59.

15. Fritz, H. G.; Aichholzer, W.; Seidenstücker, T.; Wiedmann, B. *Starch/Stärke* **1995**, *47*, 475-491.

16. Lim, S-T.; Jane, J-L. *Starch/Stärke* **1996**, *48*, 444-448.

17. Stein, T. M.; Greene, R. V. *Starch/Stärke* **1997**, *49*, 245-249.

18. Aichholzer, W.; Fritz, H.-G. *Starch/Stärke* **1996**, *48*, 434-444.

19. Anonymus, DIN-Taschenbuch 18: Kunststoffe. Mechanische und thermische Eigenschaften; Beuth Verlag GmbH: Berlin, Köln, **1988**, 180-187.

20. Debye, P. *Appl. Phys.* **1944**, *15*, 338.

21. Zimm, B. H. *J. Chem. Phys.* **1945**, *13*, 141.

22. Wyatt, P. H. *J. Analytica Chimica Acta* **1993**, *272*, 1-40.

23. Anonymus, DIN-Taschenbuch 48: Kunststoffe. Chemische und optische Gebrauchseigenschaften, Verarbeitungseigenschaften; Beuth-Verlag, Hamburg, **1988**; pp. 226-233.

24. Saechting, H.; Zebrowski, W. Kunststoff-Taschenbuch, Carl Hanser Verlag: München, Wien, **1989**; pp. 88-92.

25. Funke, U.; Bergthaller, W.; Lindhauer, M.G. Polymer Degradation and Stability **1998** (in press).

26. Kuhn, M.; Elsner, G.; Gräber, S. *Starch/Stärke* **1989**, *41*, 467-471.

27.   Bindzus, W. Untersuchung des Extrusionsverhaltens von Stärkemischungen unter Verwendung der in-line bestimmten Schubspannung der plastifizierten Massen als den Prozeß und die Produkteigenschaften charakterisierender Verfahrensparameter; PhD-Thesis, D 83, FB 15, No. 097; Technical University Berlin: Berlin, 1997; pp. 199.

[1] Publication No.: 6916 of the Federal Centre of Cereal, Potato and Lipid Research.

Chapter 3

Nanocomposite Materials of Thermoplastic Polymers Reinforced by Polysaccharide

A. Dufresne and J. Y. Cavaillé[1]

Centre de Recherches sur les Macromolécules Végétales (CERMAV-CNRS),
Université Joseph Fourier, BP 53, 38041 Grenoble Cedex 9, France

Biosynthesis of starch leads to semicrystalline systems, in which highly crystalline domains are embedded in an amorphous matrix. The average size of such microcrystals is a few tens of nm. In the case of cellulose, microfibrils are biosynthesized and deposited in a continuous fashion. This mode of biogenesis can lead to crystalline microfibrils with a high aspect ratio, almost defect free, and axial mechanical properties approaching those of perfect crystals. The use of such microcrystals, as reinforcing fillers in thermoplastic matrices leads to nanocomposite materials. In this work, we will focus on different materials, and we will see that in addition to some practical applications, their study can help to understand some physical properties as geometric and mechanical percolation effects.

Natural polysaccharide fillers are gaining attention as a reinforcing phase in thermoplastic matrices (1-3). Low density, highly reduced wear of processing machinery and a relatively reactive surface may be mentioned as attractive properties, together with abundance and low price. Moreover, the recycling by combustion of polysaccharide filled composites is easier in comparison with inorganic filled systems. Nevertheless, such fillers are used only to a limited extent in industrial practice, which may be explained by difficulties in achieving acceptable dispersion levels.

An alternative way to palliate this restriction consists of using a latex to form the matrix. Indeed, it is well known that emulsion polymerization can lead to materials easily processed, either by film casting techniques (water evaporation for paint applications for example) or by freeze drying (or more simply by flocculation) followed by a classical extrusion process. As a matter of fact, different monomers can be statistically copolymerized to adjust the glass-rubber transition temperature (4).

[1]Current address: GEMPPM, UMR CNRS-INSA #55, INSA Lyon, F-69621 Villeurbanne Cedex, France.

More generally, it is also possible to mix different types of water suspensions, including some polymer latices and organic or inorganic stabilized suspensions. In this work, various polysaccharide fillers are used and the behavior of the resulting nanocomposite materials is analyzed as a function of the aspect ratio (L/d, L being the length and d the aspect ratio).

Experimental

Transmission Electron Microscopy. All observations were made with a Philips CM200 electron microscope. A drop of a dilute suspension of microcrystalline polysaccharide filler was deposited and allowed to dry on a carbon coated grid.

Starch Microcrystals. Starch microcrystals, which are mainly formed of amylose, were prepared by acid hydrolysis of the amorphous domains of potato starch granules obtained from Sigma (Ref. 9005-25-8) as previously reported (5). They occur as a finely divided white powder, insoluble in water. Starch granules were first hydrolyzed with hydrochloric acid. This suspension (5% w/w of solid component in water) was stored at 35°C for 15 days. This period of time allows removal of the amorphous zones without damaging the crystalline zones. The suspension was stirred every day in order to ensure the homogeneity of the suspension. It was then diluted with an equal volume of distilled water and washed by successive centrifugation (4,000 trs/min) until acid free. The dispersion of microcrystals was completed by a further 3 min ultrasonic treatment (B12 Branson sonifier). The solid fraction of this aqueous suspension was determined to be around 1.5%. A typical electron micrograph obtained from the dilute suspension of hydrolyzed starch is shown in Figure 1a. It consists of starch fragments which have a homogeneous distribution in size. Each fragment contains associated microcrystals which are not clearly identified. The microcrystals are a few tens of nm in diameter and their aspect ratio is around 1.

Cellulose Microcrystals. Two sources of cellulose whiskers were used. One source was whiskers from tunicin, an animal cellulose whose microfibrils are particularly well organized and crystallized. The second source was from wheat straw cellulose.

Cellulose Whiskers from Tunicin. Suspensions of tunicin whiskers in water were prepared as described elsewhere (6,7). A batch of edible-grade tunicate (*Microcosmus fulcatus*), from the Mediterranean, was obtained from a local fish shop. After anesthetizing with chloroform, the animals were gutted and their tunic was cut into small fragments that were deproteinized by three successive bleaching treatments, following the method of Wise *et al.* (8). The bleached mantles (the "tunicin") were then disintegrated in water with a Waring blender (at a concentration of 5% by weight). The resulting aqueous tunicin suspension was mixed with H_2SO_4 to reach a final acid/water concentration of 55% (weight fraction). Hydrolysis conditions were 60°C for 20 min under strong stirring. A dispersion of cellulose whiskers resulted. After sonication, the

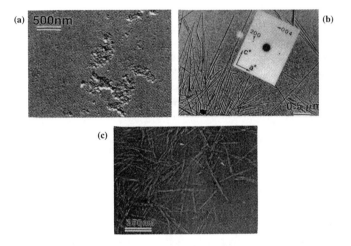

Figure 1. Transmission electron micrographs from a dilute suspension of : (a) hydrolyzed potato starch (Reproduced with permission from ref. 5. Copyright 1996 American Chemical Society), (b) hydrolyzed tunicin (inset: typical electron diffractogram recorded on one microcrystal), and (c) hydrolyzed wheat straw cellulose.

suspension was neutralized and washed by dialysis. It did not sediment or flocculate as a consequence of surface sulfate groups created during the sulfuric acid treatment (9). When concentrated by evaporation, the suspensions displayed typical liquid crystal characteristics (9,10).

A typical preparation of tunicin crystals is shown in Figure 1b. This sample consists of parallel rods with lengths ranging from 100 nm to several micrometers and widths on the order of 10-20 nm. The aspect ratio of these whiskers is, therefore, around 100. Upon testing by a microelectron diffraction technique (inset), each element gave a spot diffractogram that corresponded to a section of the reciprocal lattice of cellulose Iβ (here the a*c* section) and persisted when the electron probe was scanned along a given rod. The diffractogram indicated that the cellulose chain axis is along the long dimension of the rods. Each rod is, therefore, a whisker-like tunicin crystal with no apparent defect.

Cellulose Whiskers from Wheat Straw. The starting raw material used to prepare microcrystalline wheat straw cellulose was steam exploded wheat straw. The wheat straw was first hydrolyzed in an aqueous solution of sodium sulfite (5% v/v). Material was placed in a steam explosion reactor, developed at CERMAV (11), for 4 min at 210°C under a vapor pressure of 19.5 bars. The pressure was then rapidly released. Next, the material was washed with cold water to remove solubles. The steam exploded straw cellulose (20 wt% solids in water) was washed 6 times in boiling 2% sodium hydroxide for 4 hours under mechanical stirring. Cellulose was then bleached with a chlorite solution to a high brightness level.

Stable colloidal dispersions of straw cellulose were prepared according to a previously described method (12). The aqueous suspension of cellulose (5 wt% of filtered cellulose) was first hydrolyzed with sulfuric acid. The microcrystalline suspension was then dialyzed against distilled water until neutrality was obtained. The dispersion of cellulose microcrystals was completed by a 3 min ultrasonic treatment (B12 Branson sonifier).

A typical electron micrograph obtained from a dilute suspension of hydrolyzed straw cellulose is shown in Figure 1c. The dried suspension was negatively stained with uranyl acetate. This treatment reveals individual microcrystals (indicated by arrows) and associated microcrystals. Individual microcrystals are rods of 150 to 300 nm in length with widths close to 5 nm. This latter value corresponds to a close packing of about 40 chains of cellulose. The aspect ratio of these whiskers is, therefore, around 45.

Polysaccharide Filler/Latex Composites. In order to process nanocomposite materials with a good level of dispersion, the thermoplastic matrix used was an aqueous suspension of polymer, i.e. a latex. The latex used for the matrix was obtained by copolymerization of styrene (34 wt%) with butyl acrylate (64 wt%) and was provided by Elf-Atochem (Serquigny, France). It contains 1% acrylic acid and 1% acrylamide. It will be referred to as poly(S-co-BuA). The aqueous suspension contained spherical particles with an average diameter around 150 nm and had a 50 wt% solid fraction. The glass-rubber transition temperature (Tg) of the copolymer was around 0°C.

Various amounts of the colloidal microcrystalline starch dispersion or cellulose whisker suspension were mixed with the suspension of latex to adjust film composition. After stirring, the preparations were cast and evaporated (samples labeled E). In order to test the effect of the processing technique on the mechanical behavior of the nanocomposite materials, films were also processed by freeze-drying and molding the preparations using wheat straw cellulose whiskers as filler (samples labeled P). Hot-pressing was performed with a Carver Laboratory Press at 140 bar for 30 min at 90°C.

Dynamic Mechanical Analysis. Dynamic mechanical tests were carried out in the glass-rubber transition temperature range of poly(S-co-BuA) with two kinds of apparatus. The first one was an RSA2 spectrometer from Rheometrics, working in the tensile mode. This setup measured the complex tensile modulus E^*; i.e., the storage component E' and the loss component E''. Measurements were performed under isochronal conditions at 1 Hz and temperature was varied by steps of 3 K. The second apparatus used was a mechanical spectrometer (Mecanalyseur) from METRAVIB SA (Ecully, France). The device consists of a forced oscillation pendulum. It provides the real (G') and imaginary (G'') parts of the shear modulus and the internal friction coefficient (tan $\phi = G''/G'$) as a function of frequency and/or temperature. Temperature scans were performed in the range 200 to 500 K at a fixed frequency (0.1 Hz).

Water Uptake. Polysaccharides are hygroscopic material and water diffusion behavior will depend on composition. The use of hydrophilic fillers as a dispersed phase in a hydrophobic matrix protects against moisture. However, if the adhesion level between the filler and the matrix is not good, diffusion pathways can exist or can be created by mechanical solicitation. The existence of such pathways is also related to filler connection and, therefore, to its percolation threshold. The kinetics of absorption were determined for all the microcrystalline starch/poly(S-co-BuA) compositions. The specimens used to analyze the water absorption were thin rectangular strips with dimensions of 10 mm × 10 mm × 1 mm. The films were considered to be thin enough such that molecular diffusion was one-dimensional.

Samples were first dried for one night at 70°C. After cooling to room temperature and weighing the samples, they were immersed in distilled water. The water temperature was maintained at 25±0.5°C. Samples were then periodically removed from the water, gently blotted with tissue paper to remove the excess water on the surface, and immediately weighed on a four-digit analytical balance. In order to evaluate the water uptake of the starch microcrystal filled composite materials, the increase in weight was measured after various times of exposure to water. The weight percentage increase was calculated as follows :

$$\text{water uptake (\%)} = \frac{M_t - M_o}{M_o} \times 100 \tag{1}$$

where M_t and M_o are the weights after t min exposure to water and before exposure to water, respectively. The mass of water absorbed at time t, (M_t-M_o), can be expressed at short times (corresponding to (M_t-M_o)/$M_\infty \leq 0.5$) as reported by Comyn (*13*):

$$\frac{M_t - M_o}{M_\infty} = \frac{2}{L}\left(\frac{D}{\pi}\right)^{1/2} t^{1/2} \tag{2}$$

where M_∞ is the mass absorbed at equilibrium, 2L is the thickness of the film and D is the diffusion coefficient.

Results and Discussion

Starch Microcrystal Based Nanocomposites. The mechanical behavior and the water uptake of the starch microcrystal/poly(S-co-BuA) films were analyzed as a function of the material composition (pure matrix to 60 wt% starch microcrystal filled composites).

Mechanical Behavior. Dynamic mechanical tests were performed in the glass-rubber transition range of the matrix (*5*, Dufresne, A.; Cavaillé, J.Y. *J. Polym. Sci., Part B*, in press). The plot of log(E'/Pa) (storage tensile modulus) versus temperature at 1 Hz is displayed in Figure 2. The curve corresponding to the pure matrix (0% filled composite) is typical of thermoplastic behavior. For temperatures below the Tg, the copolymer is in the glassy state. The modulus decreases slightly with temperature, but remains roughly constant (around 2 GPa). Then, a rapid decrease in the elastic tensile modulus, more than three (3) orders of magnitude, is observed, corresponding to the glass-rubber transition. In the terminal zone, the elastic tensile modulus becomes lower with temperature and can not be experimentally measured.

At the utilized filler concentrations, there are on the order of 200,000 to 400,000 cm^2 of filler surfaces/cm^3 of material. Therefore, neglecting the likely agglomeration of the filler, the average interparticle distance is on the order of the filler diameter. This should have an effect on conformational properties and will certainly result in significant increases in the Tg of the matrix. However, the observation that should be emphasized is that the temperature of the modulus drop associated with the Tg remains almost constant, whatever the concentration of filler (*5*).

For temperatures below the Tg, the composite modulus increases up to 4 GPa for the 60% filled material. However, the exact determination of the glassy modulus depends on the precise knowledge of the sample dimensions. In this case, at room temperature, the composite films were soft and it was difficult to determine precisely sample thickness. In order to minimize this, the samples were frozen in liquid nitrogen prior to experimentation to accurately measure their dimensions.

Above the Tg, a greater increase in the composite modulus is observed with increasing volumes of microcrystalline starch. For instance, the relaxed modulus at the Tg + 50°C (~ 325 K) of a film containing only 30% starch is 100-fold higher than that

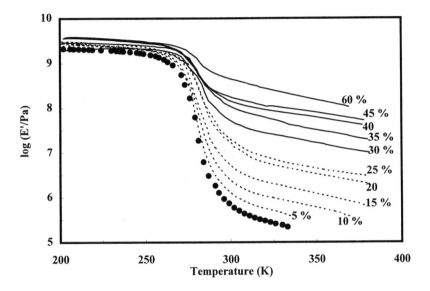

Figure 2. Storage tensile modulus E' versus temperature at 1 Hz for composites reinforced by weight fractions of starch microcrystals from 0 to 60 wt%: (●) 0%, (---) low filler content, and (——) high filler content.

of the matrix. For the 60% microcrystalline cellulose filled composite, the relaxed modulus is increased about 1000-fold. The experimental data show two sets of relaxed moduli, depending on the material composition. The first one (see dashed lines in Figure 2) corresponds to low starch content composites (from 5 to 25 wt%), and the second one (see continuous line in Figure 2) to highly filled material (from 30 up to 60 wt%). This critical value of 30 wt% more or less corresponds to a filler volume fraction of 20%, which in turn, corresponds to the percolation threshold of isoradius spherical particle filled systems. Therefore, a connection of starch particles or a geometric percolation effect occurs as the filler content approaches 20 vol%.

Starch microcrystals bring a great reinforcing effect, especially at temperatures higher than the Tg of the synthetic matrix (5). They can, therefore, be used as an economic and environmentally friendly particulate filler, and can be useful for the processing of stiff small-sized wares. Classical models for polymers comprising nearly spherical particles, based on a mean field approach as generalized by the Kerner equation according to Lewis (14) and Nielsen (15), were used to predict this behavior (5). The Kerner equation and its modifications were derived for spherical particles in a linear elastic matrix. However, this does not imply that it may be applied only below the Tg. Indeed, it is well known that the behavior of polymers is not purely elastic, but rather viscoelastic. For this reason, it is appropriate to modify the relationships for elasticity by introducing viscoelastic moduli; i.e., their complex form $E*(i\omega,T)$. However, at the temperature at which calculations were performed (325 K), this is not necessary because the storage modulus E' is about 10 times higher than the loss modulus E''. Therefore, the error made in taking E as E' is around 10%. It was observed that the calculated moduli from this model do not fit the experimental moduli of the real systems (5). This discrepancy is due to the morphology of these nanocomposite systems, and was discussed in light of aggregate formation and percolation concepts (Dufresne, A.; Cavaillé, J.Y. *J. Polym. Sci., Part B*, in press).

Water Uptake. The maximum swelling rate, or maximum relative water uptake, of starch microcrystal filled poly(S-co-BuA) was analyzed versus starch content (Dufresne, A.; Cavaillé, J.Y. *J. Polym. Sci., Part B*, in press). Water sensitivity increases linearly with the starch content. This can be easily understood because water uptake reflects an equilibrium state. If the diffusion coefficient of water, D, as calculated from equation 2, is reported as a function of the material composition, two well separated zones are displayed (see Figure 3). Above the percolation threshold (~ 20 vol%), the percolation of starch microcrystals leads to a more abrupt evolution of D versus starch content. The full line corresponds to the prediction of this behavior by a three branch series-parallel model, including the percolation concept which was developed elsewhere (Dufresne, A.; Cavaillé, J.Y. *J. Polym. Sci., Part B*, in press). The discrepancy observed between experimental data and the predicted curve in the high starch content range probably originates from the fact that the value of D used for starch in the model is that observed for native starch, which is certainly higher than that of microcrystalline starch.

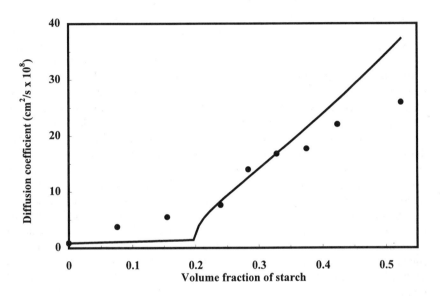

Figure 3. Variation of the diffusion coefficient versus starch content: (●) experimental data, and (—) predicted data from the series-parallel model including the percolation concept.

Cellulose Whisker Based Nanocomposites. The mechanical behavior of the cellulose whisker/poly(S-co-BuA) films as analyzed as a function of the material composition utilizing two sources of cellulose whiskers.

Cellulose Whiskers from Tunicin. Figure 4 shows the plot of log(G'/Pa) versus temperature for various tunicin whisker compositions ranging from 0 to 14 wt%. When reinforced by a small percentage of whiskers, the polymer films show improved mechanical properties which are particularly striking when the films were heated above the glass transition of the polymer. For temperatures below the Tg, the difference between the elastic shear modulus of the cellulose whiskers (50 GPa) and that of the matrix (1 GPa), is not high enough for an appreciable reinforcement effect. However, this effect exists and is well predicted by models based on a mean-field approach.

Above the Tg, a greater increase in the composite modulus is observed with increasing volume of cellulose, and, therefore, the drop in G value is dramatically reduced. For instance, the relaxed modulus of a film containing only 6 wt% of whiskers is 1000-fold higher than that of the matrix. Moreover, the reinforced films behave like rubber as their modulus value stayed constant over a wide temperature range. This is illustrated in Figure 4. The relaxed modulus of a 6 wt% filled nanocomposite remained near 0.1 GPa up to 500 K, a temperature at which cellulose starts to decompose.

The variation of the relaxed shear modulus G, taken at 325 K (i.e., 50 K above the Tg of the matrix), is plotted as a function of the whisker content in Figure 5. The experimental data (full circles) are much higher than those predicted by a classical mean-field mechanical model developed for short fiber composites (dashed line). In such an approach, following Halpin and Kardos (16), the modulus and the geometry of the fibers are accounted for, but one assumes that there is no interaction between the fibers. In particular, the mean-field approach is based on the concept that a material made of short fibers, homogeneously dispersed in a continuous matrix, is mechanically equivalent to a superposition of four plies. Within each ply, the fibers are parallel to one another and the mutual orientation of the plies is 0, +45, +90 and -45°. The mechanical properties of each ply are derived from the micromechanic equations of Halpin-Tsai (17).

In order to explain the unusually high G values of the reinforced films, one needs to invoke (i) a strong interaction between the whiskers and (ii) a percolation effect. The influence of such effects on the mechanical properties of the films can be calculated following the method of Ouali et al. (18) in their adaptation of the percolation concept to the classical parallel-series model of Takayanagi et al. (19). The shear modulus of the composite is then given by:

$$G = \frac{(1 - 2\psi + \psi v_R)G_S G_R + (1 - v_R)\psi G_R^2}{(1 - v_R)G_R + (v_R - \psi)G_S} \qquad (3)$$

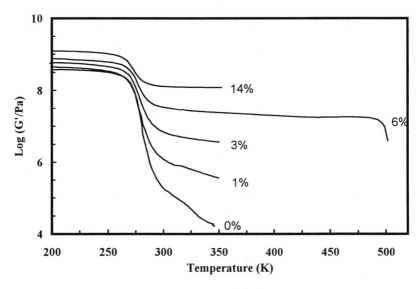

Figure 4. Storage shear modulus G' versus temperature at 0.1 Hz for composites reinforced by weight fractions of tunicin whiskers from 0 to 14%.

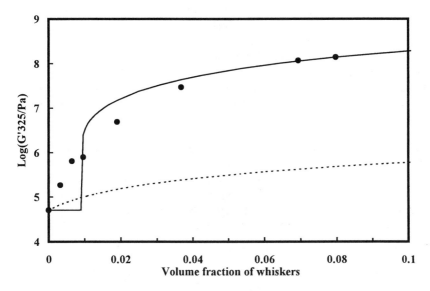

Figure 5. Plot of the logarithm of the storage shear modulus at 325 K as a function of the volume fraction of tunicin whiskers. Comparison between the experimental data (●) and calculated data with two different mechanical models: a mean-field model (dashed line) and a percolation model (continuous line).

where the subscripts S and R refer to the soft and rigid phases, respectively, and v_R is the volume fraction of whiskers. In the Takayanagi *et al.* model (*19*), ψ is an adjustable parameter, whereas in the Ouali *et al.* prediction (*18*), which was used here, ψ corresponds to the volume fraction of the percolating rigid phase. It can be calculated with a simple prediction based on the percolation concept (*18*):

$$\psi = 0 \qquad\qquad\qquad \text{for } v_R < v_{Rc}$$

$$\psi = v_R \left(\frac{v_R - v_{Rc}}{1 - v_{Rc}} \right)^b \qquad \text{for } v_R \geq v_{Rc} \tag{4}$$

where b is the percolation exponent and v_{Rc} is the percolation threshold, which is close to 1 vol% (i.e. 1.5 wt%) as determined by numerical calculation for sticks with an aspect ratio equal to 100 (*6*). According to several studies based on the percolation concepts (*20,21*), b takes the value of 0.4 in a three-dimensional system.

The calculated curve based on the percolation theory is reported as a solid line in Figure 5. It follows the dashed line up to a whiskers volume fraction of 1%. The discrepancy observed between the calculated curve based on the percolation approach and the experimental data at low filler content is probably due to the fact that the prediction does not account for the length distribution of the whiskers. Above this critical percentage, the solid line precisely fits the experimental G values without adjustable parameters. The role of percolation of cellulose fibers in paper making is well documented (*22,23*). It is, in particular, established that the high mechanical properties of a paper sheet result from the hydrogen-bonding forces that hold the percolating network of the fibers. This hydrogen-bonded system is responsible for the unusual mechanical properties of the cellulose whisker based composites when the percolation threshold is reached. Moreover, this whisker network is also responsible for the stabilization of G over a large temperature range above the Tg. It will be only when the whiskers start to decompose at around 500 K that this stabilization will disappear, inducing a catastrophic decrease of the mechanical behavior. Therefore, although mainly phenomenological, the model based on percolation concepts is able to take into account the microstructural parameters of the composites. Finite element simulations have reinforced the hypothesis that the geometrical percolation of the whiskers has been accounted for in these systems, but they also proved the important role of joints between percolating whiskers. In that regard, stiff links, due to many hydrogen bonds, are thought to exist between connected whiskers, making the whole network very rigid (*24*).

Cellulose Whiskers from Wheat Straw. The mechanical behavior of wheat straw cellulose whisker/poly(S-co-BuA) films was analyzed with a spectrometer (RSA2) from Rheometrics working in the tensile mode. When nanocomposite structures were obtained by freeze-drying and molding the mixture of cellulose and latex (samples P), the mechanical properties are substantially improved by increasing the amount of filler (*12*). Whiskers bring a great reinforcing effect at temperatures higher than the glass transition temperature of the matrix and improve the thermal stability of the composite. However, due to the lower axial aspect ratio of these whiskers in comparison with the tunicin whiskers, the reinforcement is lower. For instance, the relaxed modulus is more than a thousand times higher than that of the matrix for a film containing 30 wt% wheat straw whiskers. Figure 6 displays the relaxed modulus appraised at 325 K (~ Tg + 50K). Experimental data for the freeze-dried and hot pressed samples correspond to the full circles.

Classical models for short fiber composites based on a mean-field approach (Halpin-Kardos model) do not explain this reinforcing effect. It was observed that calculations overestimate the glassy modulus, whereas the rubbery modulus was underestimated. To fit the experimental data over the whole temperature range, interactions between the microcrystals, their topological arrangement, and the probable formation of whisker clusters within the thermoplastic matrix have to be taken into account (*12*). We have fitted the experimental moduli via a Halpin-Kardos-like equation to obtain the best interpolated curve for the entire composition and temperature ranges. From trial and error, it appears that the aspect ratio should be 450 to fit the experimental data (see dashed line in Figure 6). This means that wheat straw cellulose whiskers act as fibers (~ 10 times larger than the real ones), due to strong interactions between whiskers.

The evolution of the experimental relaxed moduli versus composition for the cast and evaporated composite films (samples E) are reported as open circles in Figure 6. It is clear from this plot that this processing technique leads to better mechanical properties than those of materials prepared by freeze-drying and hot-pressing the same mixture. This phenomenon can ensue from two origins:

  (i) a strong interaction between the whiskers leading to the formation of a whisker network within the thermoplastic matrix,
  (ii) a heterogeneity within the thickness of the composite due to the processing itself, via the evaporation step leading to the sedimentation of the cellulose whiskers, which facilitates network formation (point i).

In order to verify the second, scanning electron microscopy (SEM) was used to characterize the morphology of whisker filled nanocomposite films E (*25*). It was shown that the top face of the sample is lower in cellulose content than the bottom face. This gradient of whisker concentration is probably induced by the processing technique since it is not present in freeze-dried and hot-pressed materials. When the filler content increases, the concentration difference in the sample thickness is not so obvious. To validate this sedimentation phenomenon, wide angle X-ray scattering (WAXS) was used (*25*). Due to the high crystallinity level of the cellulose whiskers and the amorphous state of the polymeric matrix used, it is possible to determine the sedimentation of the filler by simply comparing the diffracted X-ray beam from the two faces of the sample. Differences are in good agreement with filler sedimentation.

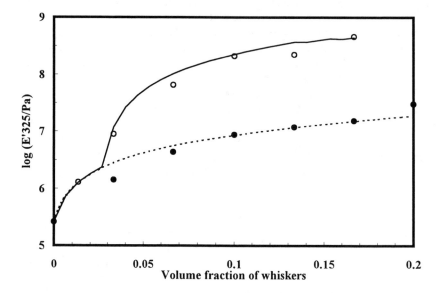

Figure 6. Plot of the logarithm of the storage shear modulus at 325 K as a function of the volume fraction of wheat straw cellulose whiskers. Comparison between the experimental data of freeze-dried and hot pressed (●) and cast and evaporated samples (O), and calculated data with two different mechanical models: a mean-field model (dashed line, L/d = 450) and a multilayered model (continuous line).

The behavior of these composite materials are well described by using a multilayered model consisting of layers parallel to the film surface and described by Dufresne *et al.* (*25*). Assuming a concentration gradient in the film thickness, the modulus of the specimen was calculated from the modulus of each layer. The modulus of each layer was predicted from a model based either on a mean field approach (Halpin-Kardos model), or on a percolation approach dependent on the volume fraction of whiskers in a given layer with respect to the critical volume fraction at the percolation threshold. The very large reinforcing effect reported for the cast and evaporated materials is, therefore, attributed on one hand to the sedimentation of the filler during the evaporation step and on the other hand to the formation of a rigid network, probably linked by hydrogen bonds. The formation of this network is assumed to be governed by a percolation mechanism. The predicted curve from this multilayered model is reported in Figure 6 (see full line), whereas open circles refer to the experimental data.

Conclusions

The present contribution, reporting work performed on the processing and behavior of new nanocomposite materials of thermoplastic polymers reinforced by polysaccharide microcrystals, is an effort aimed at providing further knowledge to a research area facing a variety of pending issues. It was shown that the reinforcing effect of a particulate filler (starch microcrystals with an aspect ratio of 1) is only due to the geometrical percolation of the filler. Increasing the aspect ratio of the filler (~ 100 for tunicin whiskers), leads to a mechanical percolation phenomenon and enhanced mechanical properties through the formation of a rigid filler network. The lower axial aspect ratio of wheat straw whiskers in comparison to the tunicin whiskers induces two main differences, namely (i) a poorer dispersion level of the filler within the synthetic polymeric matrix and (ii) a sedimentation phenomenon occurring during the evaporation step. For this latter system, mean-field effects and mechanical percolation effects coexist and the predominance of one over the other is a function of the composition of the considered layer.

The processing of fully biodegradable composites based on poly (hydroxyoctanoate) (PHO) and using starch microcrystals or cellulose whiskers as fillers, was also investigated. The results will be published shortly. The first step of this work was to purify the polymer from biomass and obtain a stable latex without degradation of the polymer (*25*). Cellulose can also be used as a microfibrillar filler, which is attractive in terms of available amounts and preparation. The mechanical behavior is then very sensitive to the degree of cellulose purification and the individual state of cellulose microfibrils (*26*). Nanocomposite materials were also prepared and characterized using potato cellulose microfibrils as the filler and potato starch as the matrix (Dufresne, A.; Vignon, M.R. *Macromolecules*, in press.).

54

Acknowledgments

Authors gratefully acknowledge coworkers who participated in results presented in this paper: Dr. H. Chanzy, Dr. V. Favier, and Dr. W. Helbert.

Literature Cited

1. Klason, C.; Kubat, J.; Strömvall, H.E. *Int. J. Polym. Mater.* **1984**, *10*, 159.
2. Zadorecki, P.; Michell, A. J. *Polym. Compos.* **1989**, *10*, 69.
3. Maldas, D.; Kokta, B. B.; Daneault, C. *Int. J. Polym. Mater.* **1989**, *12*, 297.
4. Schlund, B.; Guillot, J.; Pichot, C. *Polymer* **1989**, *60*, 1883.
5. Dufresne, A.; Cavaillé, J. Y.; Helbert, W. *Macromolecules* **1996**, *29*, 7624.
6. Favier, V.; Canova, G.R.; Cavaillé, J.Y.; Chanzy, H.; Dufresne, A.; Gauthier, C. *Polym. Adv. Tech.* **1995**, *6*, 351.
7. Favier, V.; Chanzy, H.; Cavaillé, J. Y. *Macromolecules* **1995**, *28*, 6365.
8. Wise, L. E.; Murphy, M.; D'Addiecco, A. A. *Pap Trade J.* **1946**, *122*, 35.
9. Marchessault, R. H.; Morehead, F. F.; Walter, N. M. *Nature* **1959**, *184*, 632.
10. Revol, J. F.; Bradford, H., Giasson, J.; Marchessault, R. H.; Gray, D. G. *Int. J. Biol. Macromol.* **1992**, *14*, 170.
11. Excoffier, G.; Toussaint, B.; Vignon, M. R. *Biotechnol. Bioeng.* **1991**, *38*, 1308.
12. Helbert, W; Cavaillé, J. Y., Dufresne, A. *Polym. Compos.* **1996,** *17*, 4, 604.
13. Comyn, J. *Polymer Permeability*; Comyn, J., Ed.; Elsevier Applied Science: New York, NY, **1985**.
14. Lewis, T. B.; Nielsen, L. E. *J. Appl. Mater. Sci.* **1970**, *14*, 1449.
15. Nielsen, L. E. *J. Appl. Phys.* **1970**, *41*, 4626.
16. Halpin, J. C.; Kardos, J. L. *J. Appl. Phys.* **1972**, *43*, 2235.
17. Tsai, S. W., Halpin, J. C.; Pagano, N. J. *Composite Materials Workshop;* Technomic: Stanford, CT, **1969**.
18. Ouali, N.; Cavaillé, J. Y.; Pérez, J. *J. Plast., Rubber Comp. Process. Appl.* **1991**, *16*, 55.
19. Takayanagi, M.; Uemura, S.; Minami, S. *J. Polym. Sci., Part C* **1964**, *5*, 113.
20. Stauffer, D. *Introduction to Percolation Theory*; Taylor and Francis: London and Philadelphia, **1985**.
21. de Gennes, P.-G. *Scaling Concepts in Polymer Physics*; Cornell University Press: Ithaca, NY, **1979**.
22. Batten, G. L. Jr.; Nissan, A. H. *TAPPI* **1987**, *70*, 119.
23. Nissan, A. H.; Batten, G. L. Jr. *TAPPI* **1987**, *70*, 128.
24. Favier, V.; Dendievel, R.; Canova, G. R.; Cavaillé, J. Y. *Acta Mater.* **1997**, *45*, 1557.
25. Dufresne, A.; Cavaillé, J. Y.; Helbert, W. *Polym. Compos.* **1997**, *18*, 2, 198.
26. Dufresne, A.; Cavaillé, J. Y.; Vignon, M. R. *J. Appl. Polym. Sci.* **1997**, *6*, 1185.

Chapter 4

Thermochemical Processes for Derivatization of Starches with Different Amylose Content

R. E. Wing and J. L. Willett

Plant Polymer Research Unit, National Center for Agricultural Utilization Research, Agricultural Research Service, U.S. Department of Agriculture, 1815 N. University Street, Peoria, IL 61604

Thermochemical oxidation of starches of different amylose content (0-70%) was investigated to maximize carboxyl and carbonyl content, to increase water solubility, and to give products with new and novel properties. Thermochemical processes evaluated were self-induced exothermal initiation-drum drying, jet cooking-drum drying, and drum drying using hydrogen peroxide and a ferrous-cupric catalyst. Data was presented to show product preparation, properties, and analysis. Preliminary data was presented for end-use applications in the areas of: 1) sequestration of calcium, manganese, and iron and 2) extenders in phenol-formaldehyde adhesives for plywood. Other potential applications being researched are set retardation and water reduction in concrete admixes, builders for detergents, micronutrient chelators, textile sizers, metal and rust cleaners, and cotton mercerizers.

With the increasing cost of petroleum-based products, use of natural polymer substitutes have advantages of lower cost and biodegradability, making them environmentally friendly. Starches have been oxidized with numerous oxidizing reagents for over 100 years, mainly to reduce molecular weight and to increase solubility for paper and food applications. However, all commercial products are still water insoluble. Commercial oxidized starches (*1-4*) are batch-prepared utilizing room temperature conditions and low (<3%) concentrations of oxidant (usually hypochlorite). During product isolation (filtration and water washing), some of the product is solubilized (due to molecular breakdown) and lost. While hypochlorite has been the oldest and most frequently used oxidant, other oxidants, such as permanganate, hydrogen peroxide, persulfate, periodate, and dichromate have been used. The different oxidation procedures result in variations in molecular structure and properties. The yearly consumption of oxidized starches is about 880 million kg with a predicted growth rate of 2.4% over the next decade, mainly in the adhesive area (*5*).

Maurer (*6*) reported using hydrogen peroxide (0.5%) at pH 4.5 in jet cooking starch for solubilization and use in adhesive formulations. This processing did not promote oxidation, only viscosity reduction. Ewing (*7*) obtained similar results using pH values of 6-7. Wing (*8*) reported the thermochemical oxidation of several starches by jet cooking and drum drying with several oxidants. Parovuori (*9*) studied the oxidation of potato starch with hydrogen peroxide at 40^0C with metal catalysis to yield carboxyl and carbonyl groups using acidic and alkaline reaction conditions, respectively. Kruger (*10*) obtained an oxidized insoluble starch with improved properties using manganese ion catalysis with hydrogen peroxide. Blattner and Ferrier (*11*) evaluated several metal catalysts on the oxidation of cellulose and Skibida (*12*) with several carbohydrate polymers. Floor (*13*) oxidized maltodextrins and potato starch using tungstate catalyst at low pH and high peroxide levels to yield highly degraded products of moderate carboxyl content. Whistler and Schweiger (*14*) oxidized amylopectin under ambient conditions with no catalyst over the pH range 3-11 with high peroxide levels. At pH >7, extensive carbohydrate degradation was observed, while at pH <5 little oxidation occurred. Reactive extrusion of starch has been used successfully to yield products (cationics, alkyl glucosides, carboxylates, oxidized starches, etc.) with improved reaction efficiency and solubility (*15-25*).

Thermochemical oxidation of starches at the National Center for Agricultural Utilization Research (NCAUR) at pH's of 7-11.5 with several oxidizing agents and no catalysis resulted in water soluble products with increased carboxyl content (*8*). A reactive extrusion process developed at NCAUR (*25*) of different starches with hydrogen peroxide and a ferrous-cupric catalyst gave products with increased carboxyl and carbonyl content. The present oxidation study using several thermochemical techniques evaluated starches of different amylose content with a ferrous-cupric catalyst and hydrogen peroxide to yield water soluble products of extremely high carboxyl and carbonyl content with unique properties useful in several industrial applications.

Experimental Methods

Materials. Pearl cornstarch (PCS) was supplied by Corn Products International, Engelwood Cliffs, NJ, and had an amylose content of 24%. Waxy starch (WS) was supplied by American Maize Products, Hammond, IN, and had an amylose content <1%. Amylomaize VII (AM7) was supplied by American Maize Products, Hammond, IN, and had an amylose content of 70%. Hydrogen peroxide, ferrous sulfate, and cupric sulfate were reagent grade.

Self-induced Exothermal Initiation-Drum Drying (SE-DD). Starch (200 g) was slurried in water (280 ml), catalyst (1.0 g ferrous sulfate + 0.8 g cupric sulfate) was added, and hydrogen peroxide (0-96 ml - 30%) was added. Slurries were allowed to react under ambient conditions. During the first 30 min, the pH decreased (2.7 to 1.5) and the temperature gradually increased. As the oxidation progressed, excessive foaming occurred and the temperature reached 65°C at low peroxide concentrations and 98°C at high peroxide concentrations. Solution color stayed light tan at low peroxide concentrations, but turned green at high peroxide concentrations. The reaction was over after the slurry reached its highest temperature (65-98°C). The hot solutions were immediately poured onto a double-drum drier with drum spacing of 0.005 cm at the drying temperature. The drier had 30 x 45 cm drums with a total drying surface (both drums) of 8750 cm^2. The drums were heated with 175 kg/cm^2 steam and rotated at a speed of 4 rpm. Surface temperature of the drums was 160^{0}C and the dry product came off in sheets, flakes, or powder, depending on the amount of hydrogen peroxide added. All samples were subsequently ground into a powder.

Jet Cooking-Drum Drying (JC-DD). Slurries were prepared as in SE-DD and immediately passed through a laboratory-model continuous steam-injection cooker (Penick and Ford, Ltd., Cedar Rapids, IA) at 145°C and a flow rate of 1.3 l/min. The amount of steam entering the cooker was used to regulate the dispersion temperature and the back pressure was kept constant. Final reaction mixtures were diluted ~5% due to steam condensation. The hot solutions were processed on the drum drier as previously described.

Drum Drying (DD). Slurries were prepared as in SE-DD and immediately processed on the drum drier as previously described.

Carboxyl Determination. A dry sample (1.0 g) was slurried in water (100 ml) and 0.100 N NaOH was added to keep the pH above 10. After stirring for 1 hr, the mixture was back-titrated with 0.100 N HCl to the phenolphthalein end-point. Conversion factors were determined using oxalic and citric acid as standards. Starches with no peroxide were used as a control. Samples were run in triplicate and the coefficient of variation was 1%.

Carbonyl Determination. A dry sample (1.0 g) was slurried in water (300 ml) and heated if necessary to boiling for complete solubilization. The cooled solutions were adjusted to pH 3.2 with 0.100 N HCl and 60.0 ml of a hydroxylamine hydrochloride (HMH) solution was added (HMH, 25 g; 100 ml 0.5 N NaOH diluted to 500 ml). The solutions were heated to 40^{0}C in an oven for 4 hr and titrated rapidly to pH 3.2 with 0.100 N HCl. A water sample was used as the control. Calculation: % (C=O) = 0.100 x 0.028 x (ml control-ml sample) x 100. Samples were run in triplicate and the coefficient of variation was 2%.

Solubility. Samples (10 g) were placed in water (300 ml) and the pH adjusted and maintained at 9.0 for 1 hr. After filtration, the recovered solids were oven dried and the weight determined.

Results and Discussion

Self-induced Exothermal Initiation - Drum Drying (SE-DD). The SE-DD oxidation data reported in Table I show that increasing peroxide concentration increased carboxyl and carbonyl content. The data also show as the amylose concentration is increased, slightly reduced carboxyl and carbonyl content resulted. As the amylose content increased, the data show 6 to 100% reduced solubility. Drum drying was a convenient recovery technique and didn't promote further oxidation. Also, drum drying appeared to increase product solubility, likely due to the heat (160°C) of the drums. While this oxidation method has potential to yield products for industrial application, initial foaming needs to be controlled for complete product recovery.

Jet Cooking - Drum Drying (JC-DD). The JC-DD oxidation data reported in Table II show increasing peroxide concentration increased the carboxyl and carbonyl content. The data also show increasing amylose concentration resulted in slightly increased carboxyl and carbonyl content, which is opposite to results from SE-DD data. All products were 100% water soluble, which show the increased initial temperature and turbulence from the steam caused immediate starch gelatinization and oxidation. Again, drum drying was used as a method of convenient recovery.

Table I. Effect of Peroxide Concentration on SE-DD Oxidation of Starches[a]

Starch Type	30% H_2O_2, ml	COOH, meq/g	C=O, meq/g	Solubility, %
WS	31	0.7	2.66	100
PCS	31	0.7	1.77	81
AM7	31	0.6	1.56	50
WS	62	1.5	4.99	100
PCS	62	1.3	4.30	94
AM7	62	1.0	3.60	67
WS	96	1.8	6.64	100
PCS	96	1.7	6.38	100
AM7	96	1.5	5.95	75

[a]Starch - 200 g; H_2O - 280 ml; Catalyst - 1.8 g.

Drum Drying (DD). Drum drying alone was the most efficient and economical oxidation method evaluated in this study. The DD oxidation data reported in Table III show that increasing peroxide concentration increased carboxyl and carbonyl content. Data in Table III show the DD oxidation method was not influenced by the amount of amylose in the starch. All samples using the DD oxidation procedure were recovered as a fine powder, 100% water soluble, and ready for use. A proposed scale-up for the DD oxidation procedure is shown in Figure 1. The SEM shown in Figure 2 depicts the reactive structure of a highly oxidized PCS by the DD technique.

Table II. Effect of Peroxide Concentration on JC-DD Oxidation of Starches[a]

Starch Type	30% H_2O_2, ml	COOH, meq/g	C=O, meq/g
WS	31	1.0	2.94
PCS	31	1.0	2.97
AM7	31	1.1	3.47
WS	62	1.5	5.32
PCS	62	1.5	5.65
AM7	62	1.5	6.15
WS	96	1.8	6.80
PCS	96	1.9	7.14
AM7	96	2.2	7.43

[a]Starch - 200 g; H_2O - 280 ml; Catalyst - 1.8 g.

Table III. Effect of Peroxide Concentration on DD Oxidation of Starches[a]

Starch Type	30% H_2O_2, ml	COOH, meq/g	C=O, meq/g
WS	31	0.9	2.35
PCS	31	1.0	3.14
AM7	31	1.1	3.77
WS	62	1.6	5.82
PCS	62	1.6	6.00
AM7	62	1.6	6.12
WS	96	2.2	7.93
PCS	96	2.2	7.86
AM7	96	2.3	7.94

[a]Starch - 200 g; H_2O - 280 ml; Catalyst - 1.8 g.

Sequestration of Metals. Metals (calcium, iron, magnesium, manganese, etc.) in hard water cause precipitation (metal hydroxides and oxides) problems in many industrial operations. The metals have to be kept soluble by chelation. Most industries use petrochemical-based chelants like EDTA, while others use silicates, zeolites, phosphates, etc. The afformentioned materials are not always environmentally friendly (biodegradable and non-polluting). Specifically, 1) the pulp and paper processing industry uses 450 thousand kg EDTA annually to keep iron and manganese in solution as precipitation can reduce fiber strength and discolor fiber; 2) the detergent industry uses 2 billion kg of phosphates, zeolites, silicates, etc. as detergent builders to keep calcium, magnesium, and iron from precipitating during cleaning operations; 3) the dietary fiber industry uses mainly sodium silicate for chelation of calcium, magnesium,

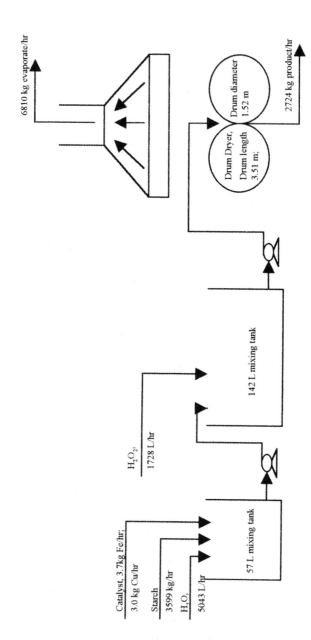

Figure 1. Production Design for Water Soluble Oxidized Starch.

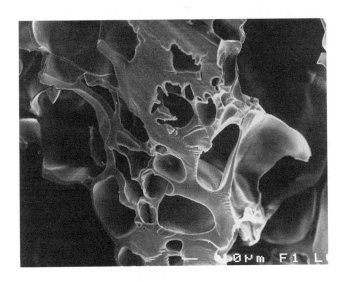

Figure 2.   SEM of Highly Active Oxidized Starch.

and iron, but the FDA is imposing strict regulations on silicon-based materials in food related products; and 4) the micronutrient industry provides at least 8 metals in chelated form (mainly EDTA) to prevent metal precipitation in soil, so metals are more available to plants. Thermochemically produced oxidized starches are being evaluated for the above mentioned applications because they are water soluble, biodegradable, and have chelating potential. The data in Table IV show chelating values of an oxidized PCS compared to EDTA at several pH's.

Table IV. Metal Chelation of Oxidized Starch Versus EDTA

Sample	Calcium Chelation at pH			Iron Chelation at pH		Manganese Chelation at pH
	12.7	9	12.7	12.7	7	12.7
	mg $CaCO_3$/g			mg Fe^{3+}/g		mg Mn^{2+}/g
Ox. PCS	400	240	20	400	125	100
EDTA-38%	<100	<100	<100	100	100	100

Adhesive Extenders. The most prominent wood adhesive is a petrochemical-based phenol-formaldehyde (P-F) resin (0.7 billion kg worldwide). Increasing cost of phenol and decreasing availability of non-environmentally friendly formaldehyde have resulted in companies searching for extenders and/or replacements for P-F resins. Some researchers have explored carbohydrates to extend P-F resins with some success and presently 7-8% wheat flour is the only commercial extender used in P-F formulations. Water soluble oxidized starches were evaluated as P-F extenders and/or substitutes in plywood adhesives. Results to date indicate 5-10% loadings of water soluble oxidized starch yield comparable adhesive strength and wood failure as shown in Table V.

Table V. Dry Shear Strength/ Wood Failure Data

Sample	Shear Strength (psi)	% Wood Failure
P-F Resin	425	89
5% Ox. PCS	401	92
10% Ox. PCS	377	90

Safety. Safety in any chemical reaction is a prerequisite. During these reactions, safety glasses and gloves were worn. All reactions were well ventilated and protective shields were used around steam. NCAUR requires SOP's and two people were present during all oxidations.

Conclusions

Several thermochemical oxidation techniques were evaluated with starches of variable amylose content to yield water soluble products possessing high carboxyl and carbonyl content. One technique (drum drying) yielded the highest activity, while being the most continuous and economical. Product evaluations for potential end-use applications as metal sequestrants and adhesive extenders were presented.

Acknowledgments

The authors wish to thank M. Redman for Figure 1, F.L. Baker for Figure 2, A.H. Conner and L.B. Dunn, Jr. for adhesive data and G. Whitehurst for chelation data.

Literature Cited

1. Roberts, H. J. In *Starch: Chemistry and Technology;* Whistler, R. L.; Paschall, E. F., Eds.; Academic Press: New York, NY, **1965**, Vol. 1; pp. 469-478.
2. Scallet, B. L. In *Starch: Chemistry and Technology*; Whistler, R. L.; Paschall, E. F., Eds.; Academic Press: New York, NY, **1967**, Vol. 2; pp. 237-251.
3. Rutenberg, M. W.; Solarek, D. In *Starch: Chemistry and Technology*; Whistler, R. L.; BeMiller, J. N.; Paschall, E. F., Eds.; Academic Press: New York, NY, **1984**; pp. 315-323.
4. Wurzburg, O. B. In *Modified Starches: Preparation and Uses*; Wurzburg, O. B., Ed.; CRC, Inc.: Boca Raton, FL, **1986**; pp. 23-28.
5. Hersch, M. In *Industry Study 645*; Freedonia Group, Inc.: Cleveland, OH, **1994**; pp. 32.
6. Maurer, H. W. *U. S. Patent 3,475,215* **1969**.
7. Ewing, F. G. *U. S. Patent* 3,539,366 **1970**.
8. Wing, R. E. *Starch* **1994**, *46*, 414-418.
9. Parovuori, P.; Hamunen, A.; Forssell, P; Autio, K.; Pautanen, K. *Starch* **1995**, *47*, 19-23.
10. Kruger, L. H. *U. S. Patent* 4,838,944 **1989**.
11. Blattner, R.; Ferrier, R. J. *Carbohydr. Res.* **1985**, *138*, 73-82.
12. Skibida, I. P.; Sakharov, A. M.; Sakharov, A. M. *Eur. Patent Appl. EP* 0548399A1 **1991**.
13. Floor, M.; Schenk, K. M.; Kieboom, A. P. G.; Van Bekkum, H. *Starch* **1989**, *41*, 303-309.
14. Whistler, R. L.; Schweiger, R. *J. Am. Chem. Soc.* **1959**, *81*, 3135-3139.
15. Marsman, J. H.; Pieters, R. T.; Janssen, L. P. B. M.; Beenackers, A. A. C. M. *Starch* **1990**, *42*, 192-196.
16. Meuser, F.; Gimmler, N.; Oeding, J. *Starch* **1990**, *42*, 330-336.
17. Chinnaswamy, R.; Hanna, M. A. *Starch* **1991**, *43*, 396-402.

18. Delle Valle, G.; Colonna, P.; Tayeb, J. *Starch* **1991**, *43*, 300-307.
19. Carr, M. E.; Kim, S.; Yoon, K. J.; Stanley, K. D. *Cereal Chem.* **1992**, *69*, 70-75.
20. Chang, Y. H.; Lii, C. Y. *J. Food Sci.* **1992**, *57*, 203-205.
21. Carr, M. E. *J. Appl. Polym. Sci.* **1994**, *54*, 1855-1861.
22. Gimmler, N.; Meuser, F. *Starch* **1994**, *46*, 268-276.
23. Tomasik, P.; Wang, Y. J.; Jane, J. L. *Starch* **1995**, *47*, 96-99.
24. Esan, M.; Brummer, T. M.; Meuser, F. *Starch* **1996**, *48*, 131-136.
25. Wing, R. E.; Willett, J. L. *Ind. Crops Prod.* **1997**, *7*, 45-52.

Chapter 5

Thermo-Mechanical Behavior of Polyethylene Reinforced by Ligno-Cellulosic Fillers

N. Bahlouli[1,5], J. Y. Cavaillé[2,6], M. García-Ramirez[3], and A. Camara[4]

[1]CIQA, Blvd.Ing.E. Reyna Hermossillo No. 140, 25100 Saltillo, Coahuila, Mexico
[2]Centre de Recherches sur les Macromolécules Végétales (CERMAV-CNRS), Université Joseph Fourier, BP 53, 38041 Grenoble
[3]Negromex, S.A. de C.V., Apdo. Postal 257, C Tampico, Tam. 89000 Mexico
[4]Universidad Autónoma de Coahuila, Blvd. V. Carranza, Saltillo, Coahuila, Mexico

Reinforced materials were prepared with (i) low linear density polyethylene (pure or mixed with 6 to 24% of a commercial poly(ethylene-co-maleic anhydride) or Bynel) and (ii) ligno-cellulosic fillers (as received or treated with 3% stearic acid). They were studied to understand the effect of adhesion between fillers and the thermoplastic matrix on thermo-mechanical behavior. The main techniques were dynamic mechanical analysis, tensile tests and Izod tests. Scanning electron microscopy was used to observe the fracture surface. It was shown that grafting of the fillers by a few % of a relatively high molecular weight macromolecule (Bynel) greatly improved adhesion, while treatment with a small molecule (stearic acid) did not lead to improvement. In addition, linear behavior was simulated using (i) a mean field equation (Kerner) and (ii) a more specific composite model (Halpin-Kardos).

Polyolefin polymers, such as polyethylene (PE) or polypropylene (PP) are widely used in the automotive industry, or for domestic applications when ductility and low price have to be combined. The decreasing number of polymers employed for such applications, resulting from the need to separate them for recycling, has emphasized the importance of PE and PP. Both are semicrystalline polymers with a glass rubber temperature (T_g) below room temperature and a melting point T_m close to 130°C for PE (depending on the precise chemical structure) and 180°C for PP. One of the consequences of the temperature range for use between their T_g and T_m is their rather poor dimension stability, and a certain ability to flow. The classical method to decrease these effects, is to fill with inorganic materials such as calcium carbonate, talc, kaolin, etc.

[5]Current address: L3MI/ULP, 15 rue du MI Lefebvre, F-67100 Strasbourg Cedex, France.
[6]Current address: GEMPPM, UMR CNRS-INSA #55, INSA Lyon, F-69621 Villeurbanne Cedex, France.

66

The counterpart of this improvement in mechanical properties is an increase of weight. In view of recycling polymeric materials, a convenient way consists of burning them as a source of energy. For this reason, the possibility of replacing inorganic fillers with low price ligno-cellulosic has to be considered. In recent years, several attempts were made to use cellulosic fillers in various synthetic matrices (J.A.Trejo-O'Reilly, J.Y. Cavaillé, A. Gandini, N.M. Belgacem, J. Adh., in press). One of the key points in processing of filled materials is the good dispersion of particles or avoiding their agglomeration, which is partly due to high surface energy in the case of ligno-cellulosic fillers. The general balance of debonding energy (W_d) between cellulose and a synthetic nonpolar polymer can be written as :

$$W_d = \gamma_{fillers} + \gamma_{polym} - 2\gamma_{fillers/polym} + W_{meca}$$

where $\gamma_{fillersl} = \gamma_{fillers}^P + \gamma_{fillers}^d$ and $\gamma_{polym} = \gamma_{polym}^d$

with $\gamma_{fillers}$ the surface energy of ligno-cellulose having a polar and dispersive component, $\gamma_{fillers/polym}$ the interfacial energy and W_{meca} the contribution coming from all inelastic processes during debonding. The surface energy can be decreased, for example, by chemical reactions on the hydroxyl groups of cellulose. If the benefit is a better dispersion, it also has the effect of decreasing W_d. One way to balance this decrease is to increase W_{meca}. This can be achieved if chemical grafting involves macromolecules fully miscible with the matrix. This is commonly done with a copolymer containing a large majority of monomer used for the matrix and a few reacting monomers having the ability of chemical interaction with the filler surface (1).

This paper will focus on low linear density polyethylene (LLDPE) filled by olive stone powder. Two different strategies were followed. They were (i) surface treatment of ligno-cellulosic particles by stearic acid, and (ii) use of a coupling agent consisting of a copolymer of ethylene and maleic anhydride.

Experimental

Raw Materials. Low linear density polyethylene (LLDPE) was provided by Mobil, referenced as MJA-202. This polymer is characterized by its low crystallinity and has a molecular weight (Mw) of 1.5E+05, and Mn of 4.5E+04. Ligno-cellulosic fillers, obtained by grinding olive stone, were provided by the Centre Technique du Papier (CTP, Grenoble, France). The irregular brown particles are composed of cellulose, lignin and hemicellulose ranging from 5 to 100 μm in size (Figure 1a).

Surface Treatment of Fillers. To improve the introduction of ligno-cellulosic fillers in the LLDPE, a superficial treatment was performed on fillers. Stearic acid was dissolved in ethyl alcohol (2 g of acid/100g of alcohol). The solution was added drop by drop to fillers previously dried at 80°C for twelve hours in vacuum and then mechanically stirred (Lightnin L1U08F Labmaster) at 400 rpm for one hour at room temperature. Stearic acid was 3% of the weight in relation with fillers. After this treatment, fillers were dried in an oven under vacuum at 105°C for 45 min.

67

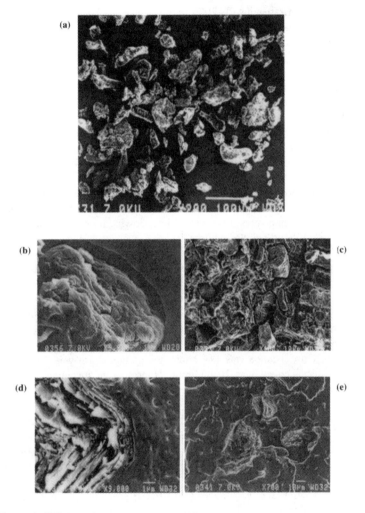

Figure 1. SEM
a : fillers,.b : LC20% x400, c : LC40% x9000, d : LC40B6 x700, e : x9000

Coupling Agent. Maleic anhydride modified polyethylene, purchased from Dupont and named CXA E409 (Bynel), was used as a coupling agent. The grafted maleic anhydride was determined by titration and the main chain was 0.21%mol (Mw=1.05E+05, Mn=3.3E+04). LLDPE and Bynel (6, 12 and 24%) were blended in a Werner & Pfleider ZSK30 twin screw extruder at 160°C and 200rpm. After that, this material was put into a Brabender PL-2000 internal mixer at 120°C and 70rpm for 12 min. This allowed all materials to have the same thermo-mechanical history. Finally, sheets were prepared by mold compression at 125°C and 69.25 kg/cm^2 for 2 min.

To melt with polyethylene, fillers were first dried under vacuum in an oven at 80°C for 12 hours in order to eliminate most of the moisture. Their incorporation into LLDPE was performed in the Brabender at 120°C and 70 rpm for 12 min. After that, sheets were prepared with the same temperature, pressure and time conditions as in the polyethylene-Bynel blends. Ligno-cellulose and treated ligno-cellulose concentration included in the polyethylene were 10, 20, 30 and 40% w/w. For treated ligno-cellulose, the same treatment and concentration described before was used. Furthermore, the same concentration of fillers was added to polyethylene-Bynel blends at the same conditions described as before.

This paper analyzes parameters dealing with the influence of the reinforcing rates and of the kind of superficial modification of fillers. To set that, different tests were performed on the composites. In this text, the following terms will be used :
LCN : LLDPE+N% of ligno-cellulosic fillers.
LCTN : LLDPE+N% of ligno-cellulosic fillers treated by stearic acid.
LCN%BM% : LLDPE+N% of ligno-cellulosic fillers containing M% of Bynel.

Tensile Tests. Tests were performed using the ASTM-D-638 standard in an United Universal Test (STM-10). Strain was measured using an United extensometer (EHE 2-1) where the maximum strain measured was 50%. Most specimens did not break under the extensometer limits. Specimen dimensions were: length-160 mm, width-12 mm and thickness-1.5 mm. Grip distance was 114.3 mm and gauge length was 50.8 mm. The speed of applied displacement was 10 mm/min. Three to five specimens were tested depending on the dispersion of experimental results.

Impact Tests. Impact tests were performed in a plastic impact tester (IZOD type) using CCF-01 CSI according to the ASTM D256-93 standard. The pendulum mass was 3.632 kg and its length was 41.526 cm. The released energy was 226.249 cm-kg$_f$. Specimen dimensions were 7 cm in length, 1.3 cm in width and 0.5 cm in thickness.

Scanning Electron Microscopy. A Jeol (JSM-6100) scanning electron microscope (SEM) was used for observing the tensile fracture surface of the samples.

Dynamic Mechanical Spectrometry. The dynamic mechanical analysis (DMA) was carried out in a bending mode using a DuPont TA 2000 thermal analyzer system equipped with the DMA 983 mechanical module. The frequency of the oscillation was 0.1 Hz and the temperature range was 30 to 120°C. Specimen dimensions were length-20 mm, width-10 mm and thickness-2 to 3 mm, depending on the rigidity of the test specimen. The tensile complex modulus E* was measured :

$$E^* = E' + i\,E'' = |E^*|\,\exp i\delta \quad \text{and} \quad \tan \delta = E''/E'$$

where E' is the storage modulus, E'' is the loss modulus and δ is the phase lag.

Results

The results obtained from tensile tests are presented in Figure 2. The Young modulus was determined from stress-strain curves (Figure 2).

It was observed that including fillers in LLDPE increased the rigidity of the composite, but decreased the maximum tensile stress. However, when Bynel was included, the modulus and the ultimate tensile stress increased in a significant way. It was also observed that 12 or 24% Bynel didn't improve composite rigidity as well as 6%, but increased maximum stress values. Treatment of fillers by stearic acid didn't effect the modulus in a significant way (Figures 3 and 4).

Figures 1b and 1c demonstrate why the maximum tensile stress decreased when fillers were included in the matrix without Bynel, which helps the compatibility between fillers and matrix. When Bynel was not used, the interface between fillers and matrix was not good. That can be readily seen from the absence of any physical contact between fillers and matrix (Figure 1b). Furthermore, fracturing the sample did not lead to breaking the fillers (Figure 1c). Adding Bynel into composites, provided a good wetting which can be observed from the absence of holes around the fillers (Figure 6a) and from the breaking of fillers during fracture (Figure 6b).

As can be seen in Figure 5, the impact energy decreased with increasing concentration of fillers, i.e., the composite became brittle when fillers were included. This phenomenon can be explained by the loss of the matrix flexibility owing to filler addition (2). The addition of Bynel improved the adhesion at the interface, but the effect of Bynel on the impact energy of composites had a maximum limit. The lowest values of impact energy were obtained for the composite containing 6% Bynel, which was found to be brittle (tensile test on Figure 2). Increasing the Bynel content (up to 12 and 24%) led to higher values of impact energy, that may be associated with the plastification of the matrix by Bynel. It is worth noting that the superficial treatment of fillers had a negligible effect on impact properties (Figure 5).

The literature concerning dynamic mechanical tests indicates that adding organic fillers into a matrix does not modify the glassy plateau (3). Therefore, the temperature range 30-120°C (rubber-like plateau) is sufficient to examine the effects of fillers and Bynel on the material. Indeed, the storage modulus, which reflects the elastic behavior of the material, increased with the concentration of fillers (Figure 6). This is in good agreement with the experimental results mentioned above. It was also noticed that Bynel had a great influence on the improvement of the dynamic mechanical modulus. Figure 7 plots the internal friction factor (tan δ) *versus* temperature. The fact that the internal friction was found to be lowest for the most rigid composite (LC40B6) supports the results described above. It can be concluded that Bynel was a good coupling agent. 6% Bynel was sufficient to improve tensile properties of the composite in question, but a higher quantity of Bynel was necessary to improve impact energy.

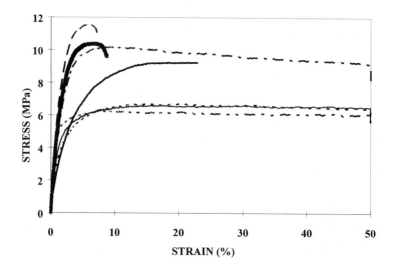

Figure 2. Tensile test : strain-stress curves.
━━ LLDPE, ┄┄ LC20%, ━━ LC30%, -..-..LC40%, ._ _ _..LC20B24, ＊ LC30B6, ┈┈ LC40B6.

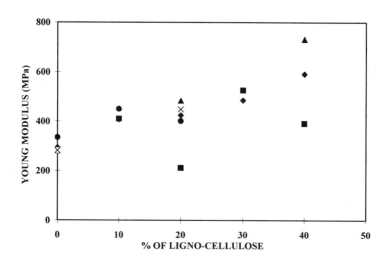

Figure 3. Influence of fillers and the different treatements on Young modulus.
◆ LCN, ■ LCTN, ▲ LCN6B, ● LCN12B, ＊ LCN24B.

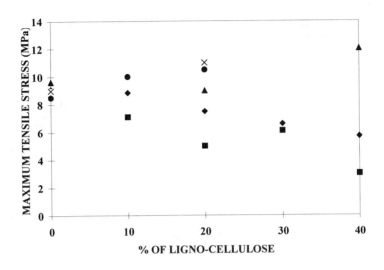

Figure 4. Influence of fillers and the different treatements on maximum tensile stress. ◆ LCN, ■ LCTN, ▲ LCN6B, ● LCN12B, ✳ LCN24B.

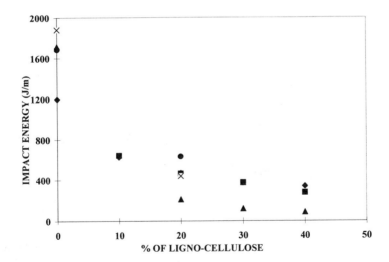

Figure 5. Influence of ligno-cellulose and treatement of fillers on the Impact energy. ◆ LCN, ■ LCTN, ▲ LCN6B, ● LCN12B, ✳ LCN24B.

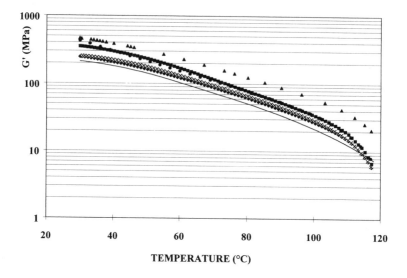

Figure 6. Storage modulus for different concentrations of fillers.
——LLDPE,   ◆ LC20%,   ● LC40%,.   ＊ LC20B24,
■ LCT40,   ▲ LC40B6.

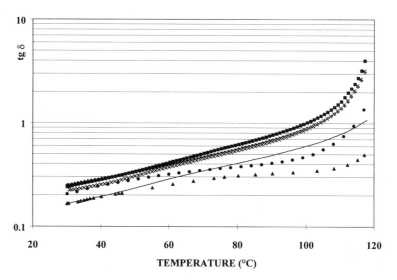

Figure 7. Damping factor for different concentrations of fillers.
——LLDPE,   ◆ LC20%,   ● LC40%,.   ＊ LC20B24,
■ LCT40,   ▲ LC40B6.

Simulation

DMA results were compared with two different models. The first one, developed by KERNER *(4)*, provides the system modulus as a function of the modulus of each phase. The basic assumptions required by KERNER's equation are the following: (a) good adhesion between phases, (b) inclusion shape is of the spherical type, (c) size and spatial distributions of fillers are at random. The equation which gives composite shear modulus G is :

$$\frac{G}{G_m} = \frac{\left[(1-v).G_m + (\alpha + v)G_f.\right]}{\left[(1+\alpha.v).G_m + \alpha.(1-v)G_f.\right]}$$

where

$$\alpha = \frac{2.(4 - 5.v_m)}{(7 - 5.v_m)}$$

G_m is shear modulus of matrix, G_f is shear modulus of fillers, v is volume fraction of fillers, and v_m is POISSON's ratio of matrix, which was held constant (0.35) *(5)*.

For the second, the Halpin-Kardos approach applied*(6)*, which is more specific for a polymeric matrix reinforced with short fibers as fillers. A heterogeneous material exhibiting reinforcement from fillers of various orientations and asymmetrical geometry may be treated as a laminated composite. In a theoretical context, the composite is considered to be mathematically equivalent to a material composed of oriented layers, the ply properties of which are specified by the volumetric and geometric properties of phases. Such a model for a random orientation of fibers is denoted as "quasiisotropic" in composite theories. The Halpin-Kardos analytical calculation was performed by considering the samples as laminates composed of four oriented plies (0°, -45°, +45°, 90°)$_s$. The modulus of each ply , E_{ii}, where *i* refers to the longitudinal (*i*=1) or transversal (*i*=2) direction of fillers, are:

$$\frac{E_{ii}}{E_m} = \frac{2E_{ii}(1+v_m)(1+\zeta_{ii}X_f) + 4\zeta_{ii}(1+v_m)^2(1-X_f)G_m}{E_{ifi}(1-X_f) + 2(\zeta_{ii} + X_f)(1+v_m)G_m}$$

$$\frac{G_{12}}{G_m} = \frac{G_f(1+\zeta_{12}X_f) + \zeta_{12}(1-X_f)G_m}{G_{ifi}(1-X_f) + (\zeta_{ii} + X_f)G_m}$$

where E_{iif} is the Young modulus in the filler direction *i* (*i*=1,2). G_f and G_m are the shear moduli of the filler and matrix, respectively, v_m is the Poisson's ratio of the matrix and X_f is the volume fraction of fillers. The aspect ratios, ζ_{ii}, are defined in the following form:

$$\xi_{11} = 2\left(\frac{L}{e}\right)$$

$$\xi_{22} = 2\left(\frac{l}{e}\right)$$

$$\xi_{12} = \left(\frac{l}{e}\right)^{\sqrt{3}}$$

where L, l and e are the length, the width and the thickness of the fibers, respectively. The shear modulus of the composite can be written as follows:

$$G = \frac{E_{11} + E_{22}(1 - v_{12})}{8(1 - v_{12}v_{21})} + \frac{G_{12}}{2}$$

with:

$$v_{12} = X_f v_f + (1 - X_f) v_m$$

$$v_{21} = v_{12} \frac{E_{22}}{E_{11}}$$

For the composite in question, the values of the parameters were :
G'_m (Pa) - experimental data, $G'_f = 5.10^9$ Pa, $E'_{11f} = 15.10^9$ Pa, $E'_{22f} = 15.10^9$ Pa, $L=l=e=50$ μm and $E'_{11f}=E''_{11f}*1000$, $E'_{22f} = E''_{22f}*1000$, $G'_f = G''_f*1000$.
The loss modulus (E'' or G'') may be considered to be negligible in comparison with the storage modulus (G', E'). Due to the thermal stability of the fillers, the mechanical properties of fillers were considered to be constant during the whole test cycle.

A good agreement was observed between the experimental DMA data and the thermograms calculated by means of the Kerner equation and the Halpin-Kardos model (Figure 8). This means that this kind of approach can be applied to simulate matrices filled with irregular shaped particles. The Halpin-Kardos approach for fillers can be used successfully with little aspect ratios. Since each of the two models implies a good wetting between fillers and matrix as a basic assumption, the fact that the best congruence between the experimental and model plots was obtained for the composite containing 6% Bynel supports experimental results described above.

Conclusions

Materials composed of fillers and a thermoplastic matrix were processed and their thermomechanical properties investigated. In order to improve the interaction between fillers and matrix, two methods were tested. A superficial treatment of ligno-cellulosic fillers with stearic acid realized no significant effect. A polymeric coupling agent was added to the composites. In this case, a significant improvement in properties was obtained. Simple mechanical modeling fit rather well with experimental data for LCBN%. SEM micrographs showed that for LLDPE+LC there was a large interfacial space between fillers and matrix. This was incompatible with the basic assumptions of the mechanical coupling models that were tested and it confirmed the strong interface adhesion for LCBN%. At very low strain, the tensile modulus increased for the three

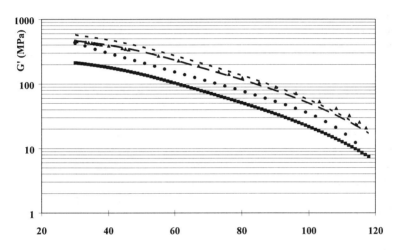

Figure 8. Simulation of storage modulus for LLDPE reinforced by 40% of fillers. ■ LLDPE, ◆ LC40%,   ▲ LC40B6, ⋯⋯⋯Kerner,⎯⎯⎯ Halpin-Kardos .

systems with increasing filler content. This increase was much higher (Figure 2) for LCBN%. This corresponded to a better interface (see Figure 1). The stress at break increased much more for LCBN%. This emphasized the role of the interface. The impact energy (and strain at break) were smallest for LCBN%. Particles were broken, but not LC and LCT. A poor interface enhanced impact properties. These effects may be understood by considering the energy W_d of debonding of matrix from filler surface. Stearic acid decreased $\gamma^P_{fillers}$ and facilitated fillers dispersion. Bynel decreased $\gamma^P_{fillers}$ as well, but increased W_{meca} and W_d. Regarding the Bynel effect on adhesion between LLDPE and filler systems, a thermomechanical approach showed that the fillers and LLDPE are compatible. Multifragmentation tests were conducted to demonstrate that the addition of Bynel to composite could enhance the load transfer between matrix and fillers and to observe if the efficacy of the load transfer would decrease when the temperature increases.

Literature Cited

1.   Felix M.; Gatenholm, P. *J. Appl. Polym. Sci* **1991**, *42*, 609.
2.   Gatenholm, P.; Kubat, J.; Mathiason, A. *J. Appl. Polym. Sci.***1992**, *45*, 1667.
3.   Favier, V.; Chanzy, H.; Cavaillé, J.Y. *Macromolecules* **1995**, *28*, 6365.
4.   Kerner, E.H., *Proc. Roy. Soc* **1956**, 69B, 808.
5.   Nielsen, L.E.; Landel, R.F. *Mechanical Properties of Polymers and Composites*; Marcel Dekker Inc., New York, **1994,** Chapter 8; pp. 463.
6.   Halpin, J.C.; Kardos, J.L. *J. Appl. Phys.* **1972**, *43*, 2235.

Chapter 6

Characterization of Plasticized and Mixed Long-Chain Fatty Cellulose Esters

P. Wang[1] and B. Y. Tao[2]

[1]Department of Chemistry, Purdue University, West Lafayette, IN 47907
[2]Department of Agricultural and Biological Engineering, Purdue University, West Lafayette, IN 47907

Fatty acid carbohydrate ester (FACE) plastics have been synthesized using a variety of mixed fatty acids esterified onto various natural carbohydrate polymer backbone chains. Physical properties can be altered by variations in fatty acid type, degree of substitution and polymer chain length. These materials can be compression molded or solvent casted, but extrusion is limited by high melt viscosities. Compounding with commercial plasticizers significantly decreased melt viscosity and improved film extrusion, but sharply reduced film tensile strength. The addition of acetyl groups increased tensile strengths of FACE polymers, while maintaining extrusion fluidity without any added plasticizer. This implies that long chain fatty acid residues act as internal plasticizers, eliminating the need for addition of synthetic plasticizing materials. This "internal" plasticization by fatty acids may also have potential importance for use in current cellulose acetate plastics, which use high levels of added plasticizers.

Long chain fatty acid esters derived from vegetable and animal fats/oils have recently seen renewed industrial interest. Driven by social interest in renewable resource utilization and environmental issues, plant lipids offer chemical structures similar to more widely used petrochemical hydrocarbons, while retaining the benefits of renewability and, in many cases, biological degradability. This interest has led to a variety of new products, ranging from fuels and lubricants to solvents, while cosmetics and surfactants are being developed (1,2).

The esterification of fatty acids with large polyhydroxyl alcohols has received relatively little study. Bradner et al. (3) noted the use of calcium and barium catalysts in the esterification of sorbitol. Konen et al. (4) described the esterification of dipentaerythritol, pentaerythritol and sorbitol with fatty acids. Current commercial interests involve the esterification of sucrose (5,6). Seino and Uchibori (7) reported using various lipases to esterify stearic, oleic and linoleic fatty acids to carbohydrate monomers and dimers. Yamane (8) discussed the engineering use of lipases for esterification of carbohydrates in micro-aqueous environments. Most recently, Harwood (9) discussed industrial lipase uses and Klibanov (10) reviewed non-aqueous enzyme catalysis. However, little has been published on the esterification of polysaccharides with long chain fatty acids.

Cellulose esters are of major commercial and scientific importance with a variety of natural and modified celluloses used industrially. The chemistry and technology of single and mixed alkyl acid esters (acetate, propionate, butyrate etc.) are well known and have been given much attention in the technical, trade and patent literature for decades (11-14). Longer chain saturated fatty acids, up to stearic acid ester of cellulose, have been previously prepared (15-17). However, very little characterization of these materials has been presented with respect to their plastic physical properties. In our previously published work, we synthesized FACE polymers from a variety of carbohydrate-lipid materials, characterized the effects of chain length and fatty acid length on plastic properties (18,19) and developed kinetic models for non-solvent synthesis (20). Here, we report work extending the physical properties of these polymers by the use of plasticizers and addition of acetyl groups to the carbohydrate backbone.

Experimental

Preparation of Cellulose. 16 Liters of 4 molar NaOH solution (1600 g of sodium hydroxide in 10 L of deionized water) was cooled to 10°C. 300 g of raw α-cellulose was added slowly (at the rate at which it wetted and sank) and kept at 10°C for 1 hr. This liquid was neutralized to about pH 4 with concentrated acetic acid. Ice was continuously added to keep the solution temperature below 25°C. The activated cellulose was collected by filtration. 20 L of deionized water was added and the suspension was slowly stirred. After washing three times, the cellulose was recovered by filtration. The batch of cellulose was dehydrated by soaking in 2000 ml of methanol for four hr. This step was repeated three times, using fresh methanol each time. This washing process was repeated 3 times, first with acetone, and then with hexane. Following filtration and vacuum drying in a desiccator for 24 hr, the cellulose was stored over desiccant at 1 mm Hg.

Preparation of Stearic Acid Chloride. 100 grams of stearic acid was added to a round-bottom-flask (500 ml volume) containing 200 ml of hexane with stirring. The flask was connected with a condenser and evacuated using water aspiration. The mixture was heated at refluxing temperature to dissolve all the stearic acid. A total of 80 g of thionyl chloride was added stepwise, in 3 ml increments. After each addition, reaction byproducts (hydrogen chloride, sulfur dioxide) were removed by vacuum. After the addition of all the thionyl chloride, the reaction mixture was refluxed for two hr. The stearoyl chloride was recovered by solvent evaporation.

Synthesis of Mixed Acid Esters. 4.0 g of hydrolyzed cellulose of molecular weight 100 kDa was added to a round-bottom-flask (250 ml) with a magnetic stir bar, followed by 40 ml of pyridine and 150 mL of 1, 4-dioxane. 24, 21, 18, 15, 12, 9, 6, 3, or 0 grams of long-chain fatty acids were first added to the flask, followed by 0, 1, 2, 3 ,4, 5, 6, 7, or 8 grams of acetyl chloride in experiments 1-9. The reaction mixtures were slowly heated to 90°C, and the reaction was conducted for 24 hr with stirring. The reaction mixtures were cooled to 40°C, 200 ml of methanol was added and mixtures were stirred for 5 min. Each mixture was filtered and the yellow powder product was washed and filtered a second time with methanol. The products were dried under reduced pressure, at room temperature for 24 hr. For high proportion acetyl chloride products, no precipitate was formed after adding methanol. In these cases, some water was added until precipitate formed, then it was filtered and dried. The yields of experiments 1-9 were 18.1, 15.8, 12.8, 10.5, 9.5, 8.7, 7.1, 6.2 and 6.0 grams.

Preparation of Mixed Ester Films by Solvent Casting and Extrusion. Pure FACE was dissolved in toluene at refluxing temperature for 3 h and good films were formed by casting onto glass plates. The mixed esters from experiments 2-6 did not dissolve in toluene and were cast from dimethylformamide. The esters from experiments 7-9 were cast from chloroform.

A mixed ester, made of three grams of acetyl chloride and fifteen grams of soy fatty acid chloride without exogenous plasticizer, was extruded at 280°F. This material was designated as FAACE (Fatty Acid and Acetyl Cellulose Ester). The properties of these films from casting and extruding were measured and the results are presented in Table VIII.

Evaluation of Plasticizers with FACE Plastics. The effects of various plasticizers with FACE polymers was measured by a variety of methods:

FACE-Plasticizer Compatibility/Solubility. The compatibility/solubility of FACE polymers with various plasticizers was measured by mixing samples of FACE with plasticizer under controlled temperatures. Compatibility/solubility was subjectively evaluated by visual observation of phase separation, swelling and/or gelation as described in the text.

Compatibility of Plasticizers with FACE Films. Toluene solutions of FACE plastics containing increasing quantities of the plasticizer were prepared and films were cast. After evaporation of the solvent, the films were examined to determine the maximum quantity of plasticizer which gave a clear film.

Effects of Plasticizers on FACE Film Physical Properties. Plasticizers with good compatibilities were mixed with FACE plastics. The mixtures were extruded by a microextruder into films. Tensile strength and elongation were measured. Correlations between percentages of plasticizers and the film properties were obtained.

Results

Plasticizer Effects with FACE.

      Compatibility/Solubility of FACE (DS=2.8) in Plasticizer. Approximately 0.3 grams of FACE (DS= 2.8) was mixed with approximately 10 g of plasticizer in test tubes. Four trials were conducted: Room temperature for 24 hr and 50°C, 100 °C, and 150°C for two hr. Then each sample was visually examined and the compatibility of the FACE in the plasticizer was rated on a scale of 1 to 5:

  0 - no swelling of FACE
  1 - very little swelling
  2 - some swelling
  3 - twice original size
  4 - three times original size
  5 - four times original size

The solubility results of FACE (DS=2.8) in plasticizers are listed in Table I.

Table I. Compatibility/Solubility of FACE (DS=2.8) with Plasticizers

Plasticizer	R. T. (24 hr)	100°C (2 hr)	150°C (2 hr)
Paraplex G-62 Soybean Oil	3	3	3
Plasthall ESO Soybean Oil	2	3	3
Plasthall 7050 NJTSRN	1	2	2
Plasthall 200 Dibutoxyl Phthalate	3	3	3
Paraplex G60 Soybean Oil	2	2	2
Dibutyl Phthalate	3	4	4
Tributyrin	1	3	5
Dipropyl Phthalate	2	3	4
Ethylene Glycol Diacetate	0	1	1
Tributyl Phosphate	4	5	4
Triacetin	0	1	0
Di(ethylene glycol) Dimethacrylate	1	3	*

* - No data taken (Sample was contaminated).
NOTE: No noticeable swelling occurred in any mixture at 50°C for two hr.

      Compatibility/solubility increased overall with increased temperature and increased significantly in the cases of tributyrin, dipropyl phthalate and di(ethylene glycol) dimethacrylate. The FACE was very soluble in tributyl phosphate and dibutyl phthalate at 100°C and 150°C, with the most soluble plasticizers being tributyrin and tributyl phosphate at 150°C, and the least soluble plasticizer being ethylene glycol diacetate.

For testing lesser substituted FACE (DS 2.1 and DS 1.2) solubilities, approximately 2.3 mg of each FACE was mixed with about 5 g of plasticizer. The resulting mixtures were evaluated at room temperature for 24 hr and at 135°C for 2 hr. Then each mixture was rated on the following scales for the compatibility/solubility of the FACE in the plasticizer:

0 - FACE not dissolved at all
1 - slightly dissolved (swelled)
2 - some/half dissolved
3 - mostly dissolved
4 - all or almost are dissolved

The solubility data are presented in Tables II and III.

Table II. Compatibility/Solubility of FACE (DS=2.1) with Plasticizers

	Compatibility/Solubility	
Plasticizer	*20°C, 24 hr*	*135°C, 2 hr*
Paraplex G-62 Soybean Oil	1	2
Plasthall ESO Soybean Oil	1	2
Plasthall 7050 NJTSRN	0	0
Plasthall 200 Dibutoxyl phthalate	1	2
Paraplex G-60 Soybean Oil	0	2
Dibutyl phthalate	4	4
Tributyrin	0	1
Ethylene Glycol Diacetate	0	1
Tributyl Phosphate	3	4
Triacetin	0	1
Di(ethylene glycol) Dimethacrylate	1	2

The data show that the most soluble plasticizers were dibutyl phthalate, tributyrin, and tributyl phosphate. The rest of the plasticizers showed little or no solubility with the low substituted FACE (DS=1.2). Also, temperature was not a major factor in the solubilities of any plasticizer.

Extrusion of FACE and Plasticized FACE. First, low substituted FACE (DS=1.2) was extruded without any plasticizer. The extrusion temperature was increased from 220°F to 340°F to see the effect of temperature on the quality of films. At 220-300°F,

Table III. Compatibility/Solubility of FACE (DS=1.2) with Plasticizers

	Solubility	
Plasticizer	Room Temp., 24 hr	135 °C, 2 hr
ParaplexG-62 Soybean Oil	0	1
Plasthall ESO Soybean Oil	0	1
Plaraplex G-60 Soybean Oil	0	1
Plasthall 7050 NJTSRN	1	1
Plasthall 2000 Dibutoxyethyl Phthalate	2	2
Dibutyl Phthalate	3	4
Tributyrin	0	1
Dipropyl Phthalate	**	**
Ethylene Glycol Diacetate	0	0
Tributyl Phosphate	3	3
Triacetin	0	0
Di(ethylene glycol) Dimethacrylate	1	1*

** Samples were contaminated

* Sample gelatinized after several days standing at room temperature.

extruded films were smooth, containing no holes or cracks, and did not curve upon exit from the die. However, films did split in the middle as they exited the die. The reason for splitting was that the flow rates along the die were not the same. The middle had the highest rate and decreased from center to both sides of the die due to low fluidity of heated FACE in the die. When the temperature was increased to 320°F, the films contained many small cracks. However, films did not split as they exited the die. This is presumed to be due to higher fluidity at 320°F. At 340°F, the films were very rough and had many small holes, but did not split in the middle. These films were then tested for tensile strength. The physical properties of the films are presented in Table IV.

Table IV. Physical Properties of FACE (DS=1.2) Films vs. Temperature

Extrusion Temp. (°F)	Tensile Strength (x 10⁵ Pa)	Elongation (%)	Color
220	32.6	4.5	light yellow
260	42.0	5.0	light yellow
300	28.1	4.0	yellow
320	21.3	5.1	yellow
340	15.0	6.2	light brown

As extrusion temperature increased from 220°F to 260°F, tensile strength increased and then decreased from a maximum value of 42.0 at 260°F. Elongation increased as temperature increased. Higher temperature helped solve the splitting problem of extruded films, but the films became rough and porous (small holes and cracks). At lower temperature (220-260ºF), extruded films were much stronger than at higher temperature, but showed serious film splitting problems because of the low fluidity at those temperatures.

A new series of extrusion tests were conducted at 260°F with increasing amounts of plasticizer (15%, 30% and 45% w/w). The plasticizer was tributyl phosphate (Bu_3P) which has the highest solubility for the FACE with DS 1.2. The results are listed in Table V.

Table V. Physical Properties of Plasticized FACE (DS=1.2) Films

Bu_3P (%)	Tensile Strength x 10^5 Pa	Elongation %	Film Color
15	4.3	6.5	yellow
30	1.7	6.1	light yellow
45	too weak to measure		very light yellow

The color of the FACE films with DS 1.2 and tributyl phosphate got lighter and lighter as the amount of plasticizer increased. The films were smooth without any small holes or cracks and felt more flexible, although elongation values were not much increased. The films became narrower than films without plasticizer and the films did not have splitting problems. From Table V, it can be seen that the tributyl phosphate plasticizer greatly reduced the tensile strengths of FACE (DS 1.2) films, although it did improve the FACE extruded film quality. Thus, using a proper plasticizer could extrude FACE at lower temperature without splitting problem and get higher quality plasticized-FACEs film (without small holes and cracks).

The intermediately substituted FACE (DS 2.1) was first extruded at temperatures of 200, 240, 260, 300 and 320ºF without plasticizer. At 200-260°F, the films split although they had smooth surfaces. When the temperature was raised to 300 and 320ºF, a lot of small cracks and holes were observed on the films and the films became rough. The physical properties of these films were measured and results are listed in Table VI.

Similar extrusion results were observed as for the case of DS 1.2 FACE. The best temperature was around 240ºF, as higher temperature caused oxidation problems which resulted in small cracks and holes and lower tensile strengths.

The FACE with DS 2.1 was mixed with tributyl phosphate which is the most soluble tested plasticizer for this FACE. The amounts of plasticizer were 15%, 30%, and 45% (w/w), and the extrusion temperature was 240ºF. The films were tested and the results are listed in Table VII.

84

Table VI. Physical Properties of Extruded FACE (DS=2.1) Films vs. Temperature

Extruded Temp. (°F)	Tensile Strength x 10⁵ Pa	Elongation (%)
200	20.12	5.1
240	20.43	5.3
260	14.56	5.6
300	9.90	4.43
320	7.43	4.26

Table VII. Physical Properties of Plasticized FACE (DS=2.1) Films

Amt. of Plasticizer (%)	Tensile Strength x 10⁵ Pa	Elongation (%)
15	1.56	5.1
30	0.78	4.7
45	too weak to measure	

From Table VII, the observation was again made that the plasticizer tributyl phosphate can help the extrusion of FACE, making the films smooth, solving curve and splitting problem, but greatly reducing tensile strengths (by about 90% of original tensile strength).

Characterization of FAACE

The total DS of mixed cellulose esters was maintained by simply substituting the appropriate amount of acetate for soy fatty acids. The physical properties of cellulose acetate and long-chain fatty acid ester plastics were measured and the results are listed in Table VIII.

Table VIII. Physical Properties of Solvent-Cast FAACE Films

Amount of acetyl chloride (g)	Amount of FA chloride (g)	Tensile Strength $x10^5$ Pa	Elongation %
0	24	27.5	30.0
1	21	33.7	25.9
2	18	47.9	20.5
3	15	90.0	16.7
4	12	134.2	13.4
5	9	183.1	10.1
6	6	228.3	6.5
7	3	287.4	6.0
8	0	341.9	5.6

From Table VIII it was observed that as the amount of acetyl chloride increased, the tensile strength increased, but elongation decreased. An extruded sample of FAACE film was made to compare its physical properties versus a solvent cast sample, using a 3 gm acetyl chloride/15 gm FA chloride material (see Table VIII). Both films had similar tensile strengths (85.1 vs. 90 x 10^5 Pa), but extruded films had a higher elongation (23.8% vs. 16.7%). Their final film tensile strength was greatly increased (about three-fold) in comparison to the FACE films.

Conclusions

Our work with FACE polymers and plasticizers has demonstrated that it is technically feasible to produce extruded plastic films. Of the dozen commercial plasticizers tested, the most compatible compounds were dibutyl phthalate, tributyrin, and tributyl phosphate over the entire range of DS values. high DS, modified soybean oils (Paraplex G-62 Soybean Oil and Plasthall ESO Soybean Oil) exhibited significant compatibility, but were ineffective at lower DS values. While the use of exogenously added commercial plasticizers, such as tributyl phosphate, can effectively plasticize FACE, the resulting films have very low tensile strengths. Using higher initial tensile strength FACE polymers, such as longer molecular weight or completely saturated fatty acids, could improve final film tensile strength.

The addition of acetate groups to high DS FACE was shown to be an effective way to create mixed acid esters with increased tensile strength. Added acetyl groups showed a ten-fold increase in film tensile strength versus the pure FACE polymers, while retaining flexibility. It is thought that the long chain fatty acid moieties function like an "internal" plasticizer for these polymers. This eliminates the need for the addition of expensive commercial plasticizers and avoids potential leaching out of the plasticizer. The use of intermediate chain length fatty acids may provide an additional means to control polymer fluidity and other properties. Our earlier work (unpublished) with mixed short chain acids also indicated that they would improve tensile strength.

By appropriately balancing the acetate-long chain fatty acid content, extruded films can be made with high tensile strength and good elasticity, without the addition of plasticizers. The ability of the long chain acids to function as an "internal" covalently linked plasticizer, eliminates the need for expensive additional plasticizers. The cost of soybean oil is significantly lower than the cost of many of the currently used plasticizers in cellulosic plastics, which can use up to 40% plasticizer. Therefore, this result is not only important for FACE production, but also may be useful in plasticizing existing high acetate cellulosic plastic products.

Acknowledgments

The authors gratefully acknowledge the support of the American Soybean Association and the Indiana Soybean Board.

Literature Cited

1. Tao, B. Y. *Chemistry & Industry* **1994**, *22*, 906-909.
2. Tao, B. Y. Industrial Applications for Soybeans, *25ᵗʰ ACHEMA*, Frankfort, Germany, **1997**.
3. Bradner, J.; Hunter, D.; Brewster, M. D.; Bonner, R. E. *Ing. Eng. Chem.* **1945**, *37*, 809-812.
4. Konen, J.; Locker, C.; Cox, R. P. *Oil and Soap* **1945**, *22*, 57-60.
5. Zeringue, J. J.; Feuge, R. O. *J. Am. Oil Chem. Soc.* **1976**, *53*(12), 719-712.
6. Rios, J. J.; Perez-Camino, M. C.; Marquez-Ruis, G.; Dobarganes, M. C. *J. Am. Oil Chem. Soc.* **1994**, *71*(4), 385-390.

7.   Seino, H.; Uchibori, T. *J. Am. Oil Chem. Soc.* **1984,** *61*(11), 1761-1765.
8.   Yamane, T. *J. Am. Oil Chem. Soc.* **1987,** *64*(12), 1657-1662.
9.   Harwood, J. *TIBS* **1989,** *14*, 125-126.
10. Klibanov, A. M. *TIBS* **1989,** *14*, 141-144.
11. Stannett, V. *Cellulose Acetate Plastics*; Temple Press Ltd.: London, **1950**.
12. Smith, R. L. *Paint Varn. Prod. Manager* **1969,** *59*, 53.
13. Fisher, J. W. In *Recent Adv. Chem. Cellulose and Starch*; Honeyman, J., Ed.; Heywood: London, England, **1959**; p. 188.
14. Keith, C. H.; Tucker, R. O. *U.S. Pat. 4,192,838*; **1980**.
15. Malm, C. J.; Fordyce, C. R. In *Cellulose and Cellulose Derivatives*; Ott, E.; Spurlin, H. M.; Grafflin, M. W., Eds.; Interscience: New York, NY, **1954**, 2nd ed., Part 2; p. 667.
16. Malm, C. J.; Mench, J. W.; Kendall, D. L.; Hiatt, G. D. *Ind. Eng. Chem.* **1951,** *43*, 684.
17. Battista, C. A.; Armstrong, A. T.; Radchenko, S. S *Polymer Preprints* **1978,** *19*, 567.
18. Wang, P.; Tao, B. Y. *J. Appl. Polymer Sci.* **1994,** *52*, 755-761.
19. Wang, P.; Tao, B. Y. *J. Environ. Polymer Degrad.* **1995,** *3*, 115-119.
20. Kwatra, H.; Caruthers, J.; Tao, B. Y. *Ind. Eng. Chem. Res.* **1993,** *31*(12), 2647-2651.

Chapter 7

Alkali Induced Polymorphic Changes of Chitin

J. Li, J.-F. Revol, and R. H. Marchessault

Department of Chemistry, Pulp and Paper Research Center, McGill University, 3420 University Street, Montreal, Quebec H3A 2A7, Canada

By treatment with 50% NaOH and subsequent washing in water, α-chitin and β-chitin undergo polymorphic transformations that have been followed by X-ray diffraction and solid state NMR. In both cases, during swelling in NaOH, the original lateral structure is destroyed and an alkali chitin complex is formed, in which the general orientation of the chitin chains remains parallel to the microfibrils axis. After washing in water, the alkali chitin from both polymorphs is converted to the α-chitin crystal structure, although the crystallinity remains poor. In the case of the β-chitin polymorph, the original parallel arrangement of the chains is changed into an antiparallel arrangement during conversion to the α-form. A mechanism of polymorphic transformation involving chain interdigitation, similar to the one for the mercerization of cellulose, is also proposed. The classical deacetylation of α-chitin in the presence of 50% alkali to produce chitosan is a permutoid reaction where alkali saturated crystallites behave as individual reactors.

Chitin, a linear polysacharride consisting of N-acetyl-D-glucosamine, is widely distributed in nature, e.g., in the shells of crustaceans and insects, and in the cell wall of bacteria (1). Chemically, chitin is similar to cellulose, differing only in the fact that chitin has an aminoacetyl group instead of hydroxyl group at C-2 (2). In nature, chitin is biosynthesized by chitin synthetases with near simultaneous crystallization to form microfibrils (3,4). The polymorphic forms of chitin in nature is determined by chitin synthetases (5). Chitin with an antiparallel chain packing is referred to as α-chitin and is that found in crab, lobster and shrimp shells (1,6). Chitin with a parallel chain packing is referred to as β-chitin and occurs in squid pen or *pogonophore* tube (7,8).

In spite of the similarity in structure with cellulose, the chemical and physical properties of α-chitin are significantly different from those of cellulose. In particular, chitin is much less reactive to many chemicals due to the peptide-like hydrogen bonds between chains (6). Chitin is insoluble in most of the organic and inorganic solvents known for cellulose (9). Many kinds of chitin derivatives, including benzyl chitin, carboxylmethyl chitin, etc., have been prepared via alkali chitin which is a soda-chitin complex (*10,11*). Preparation of alkali chitin was first reported in 1940 by Thor and Henderson, who steeped chitin in excess of 50% aqueous sodium hydroxide (*12*). Their results show that alkali chitin has similar properties to alkali cellulose in terms of reactivity with reagents of the type which react with alkali cellulose. However, compared with alkali cellulose, which has been extensively investigated, alkali chitin has received relatively little attention. It is generally accepted that native cellulose, cellulose I, which has a parallel chain structure, is converted to an antiparallel chain structure, cellulose II, after alkali treatment (*13-19*). It is expected that β-chitin, which also has a parallel packing, will behave in the same fashion. In the present study, x-ray diffraction, CP-MAS ^{13}C NMR and polarized light microscopy have been used to investigate some structural features of alkali chitin and water regenerated chitin, which is referred to as "recovered chitin" in this paper. A comparison of crystalline order and polymorphism between untreated chitin and the recovered chitin was interpreted to provide further understanding in terms of the familiar mercerization process for native cellulose.

Experiment

Materials. Crab α-chitin, purchased from Fluka, was purified as previously described (*20*). Squid β-chitin was prepared from squid pen as described below. Approximately 100 g of dry squid pen was first pulverized in a Wiley mill to pass a 20 mesh sieve. Then, the ground squid pen was deproteinized by treating with 1 N NaOH for 1 hr at 100°C, followed by demineralization using 0.5 N HCl at 100°C (1 hr). For both treatments, the ratio of solid to liquid was 1 g per 10 ml. Finally, the purified squid chitin was washed using deionized water and acetone, followed by air-drying in a fumehood.

Preparation of alkali-chitin. Alkali treated α-chitin was prepared by immersing 5 g of the purified chitin materials in a beaker with 25 ml of 10 N NaOH. To avoid deacetylation, this mixture was kept in a refrigerator at 4°C for 3 hr. For CP MAS ^{13}C NMR measurements, 1 g of alkali treated α-chitin was pressed free of excess alkali solution with filter papers. The recovered chitin was obtained by repeatedly washing 1 g of alkali chitin with deionized water, followed by filtration until the filtrant reached neutrality. Subsequently, the wet sample was pressed free of excess water with filter paper, followed by air-drying in a fumehood. Squid pen β-chitin was alkali-treated as above, followed by washing-filtration cycles. Finally, the sample was dried by filter press, followed by air-drying in a fumehood. Samples of the alkali-treated chitins were examined after they had been pressed free of excess alkali.

X-ray diffraction. All X-ray powder diffractograms were recorded using a Siemens D-5000 diffractometer with a Cu K$_\alpha$ (λ=1.5Å) radiation source equipped with a scintillator counter and a linear amplifier. A typical scan range is from 0 to 50° (2θ).

CP-MAS ^{13}C NMR. All CP-MAS ^{13}C NMR spectra of chitin samples were recorded using cross polarization magic angle spinning at 75 MHZ on a Chemagnetics CMX-300 spectrometer. Powder samples of chitin were first inserted into a 7.5 zirconia rotor. A contact time of 2 ms and recycle delay of 2 s were employed. A magic angle spinning rate of 5 kHz was used.

Polarized Light Microscopy. Alkali chitin samples were transferred to glass slides and samples were examined using a Nikon Microphot-FXA polarized optical microscope. Micrographs were recorded with the sample between crossed polars.

Results and Discussion

The X-ray powder diffractograms of original and alkali-treated α-chitin samples are shown in Figure 1. The X-ray diffractogram of the original chitin shows a typical α-chitin pattern (see Figure 1a) in keeping with the Carlström unit cell (6). Figure 1b shows that alkali chitin is noncrystalline with only a small trace of the α-chitin diffraction. When alkali chitin is neutralized by washing with water, it regains the appearance of the original α-chitin. The X-ray diffractogram of the recovered chitin shows the diffraction pattern of α-chitin, although a general line broadening and amorphous background is observed compared with the original chitin (see Figure 3a). This is due to poor lateral ordering in the recovered chitin. These results, coupled with observations with the polarizing microscope, demonstrate that immersion of chitin in a concentrated aqueous alkali solution results in swelling and the destruction of the original lateral order. Alkali ions penetrate into the crystalline lattice and break the hydrogen bonds between polymer chains. Once hydrogen bonds are broken, the polymer chains are pushed apart by water and alkali. Physically, the crystallites are swollen and, according to Thor and Henderson (12), chitin forms complexes with alkali in which an alkali chitin could contain approximately 0.75 equivalents of combined sodium, or more, per acetyl hexosamine unit, depending on the ratio of alkali to chitin and the treatment time. By comparison, the alkali cellulose complexes contain more sodium (16) and, contrary to the alkali chitin, they exhibit relatively well defined crystalline structures (16,19). However, there is still a long range order existing in the alkali chitin samples as demonstrated by the observation of a strong birefringence (not shown here), which indicates a parallel arrangement of the molecular chains. When the alkali ions in the latices are washed away, the aligned polymer chains spontaneously reorder and interchain hydrogen bonds reform. Thus, the polymer chains are reorganized into the original crystalline packing, although not as perfectly as for native α-chitin. Because this reorganization process of microfibrils is fast, some disorder or defects are unavoidable.

The results from a CP-MAS ^{13}C NMR study are consistent with those from X-ray diffraction. In particular, the spectrum of the recovered chitin is typical of the original α polymorph (see Figure 2b) (7,21). Typically, the spectrum consists of eight well defined resonances which are assigned to carbonyl groups (~174 ppm), methyl groups (~23 ppm) and six glucose carbons. The detailed assignments are given in Tanner's paper (7). For the alkali complex, the degree of order shown by the linewidth

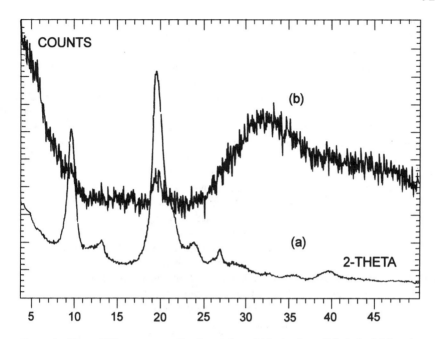

Figure 1. X-ray diffractograms of native crab α-chitin (antiparallel chains) (a) and alkali treated crab chitin (b).

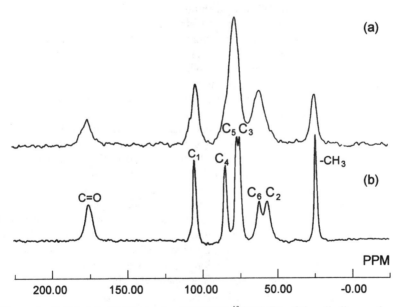

Figure 2. Solid state NMR spectra (CP-MAS ^{13}C NMR) of the alkali complex from α-chitin (a) and the recovered chitin (b).

of signals is poor (cf., Figure 2a). This confirms that a well-defined alkali chitin crystalline structure has not been achieved, but the chemical shift of the partially merged peaks leaves no doubt that an alkali chitin complex has formed.

The X-ray powder diffractograms of the squid β-chitin (not shown) indicate that β-chitin is, in general, less ordered than α-chitin. The diffraction pattern of the recovered chitin from the β-form (squid chitin) is shown in Figure 3b. It exhibits a very poor crystallinity, but the characteristics of the α-chitin can be identified and no trace of β-chitin can be detected. The CP/MAS ^{13}C NMR spectra of the original squid β-chitin and corresponding recovered chitin are shown in Figure 4. They confirm that the recovered chitin from the β-form contains a larger amount of disordered material than the recovered chitin from the α-form (see Figure 2b). The typical splitting of the C_3-C_5 resonances, which is clearly observed in Figure 2b, is only marginally evident in Figure 4a, probably due to the lack of crystallinity. Similar behaviour was observed for the recovered material from squid β-chitin after a solid state conversion into the α-polymorph by treatment in 6N HCl (7).

The fact that only a very poor crystallinity is recovered during the solid state conversion from the β-form into the α-form is an indication that the recrystallization is inhibited. On the other hand, when alkali chitin from the α-form is washed in H_2O, the return to the α-chitin crystalline structure is obviously far easier, although still not perfect (see Figure 3a). This behavior may be the direct consequence of the different chain polarity in the two polymorphs.

It is generally admitted that, in β-chitin, each microfibril contains chains of the same polarity (parallel chains) (8), whereas α-chitin contains chains of alternating polarity (antiparallel chains) (6). A reversal of the chain polarity must occur during the polymorphic transformation. When solid chitin is recovered from solutions as shown by Focher et al. (21), the reversal of the chains to form the α-structure is readily obtained by crystallization into this more favorable packing. On the other hand, during the solid state conversion, even though chitin is swollen, the molecules are not free to rotate as in solution and the conversion is severely restricted. In 1969, Rudall (22) proposed as possible mechanisms that the parallel piles of chitin chains, which are present in the β-form, separate in alkali and recombine with nearest antiparallel neighbors to give the α-chitin structure upon washing in water.

Such a concept has also been considered and discussed for the mechanism of mercerization of cellulose, in which a similar chain reversal occurs to convert, in the solid state, the parallel packing of cellulose I into the antiparallel packing of cellulose II (13-19). More specifically, the following has been proposed: Native cellulose microfibrils contain chains of the same polarity, but they point alternately up and down in the cell wall (23). Two neighboring microfibrils of opposite polarity may coalesce during swelling in alkali and allow the chains to adopt the antiparallel arrangement needed to form cellulose II. Crystallographic studies seem to indicate that an antiparallel arrangement is already present in the alkali cellulose (16). Similarly to cellulose, the solid state phase transformation of β-chitin to α-chitin can be explained by the mingling of chains from adjacent and antiparallel β-chitin microfibrils to form α-chitin crystals of antiparallel chains. Perhaps more important than the β→α conversion mechanism, is the role played by the alkali intermediate in the deacetylation

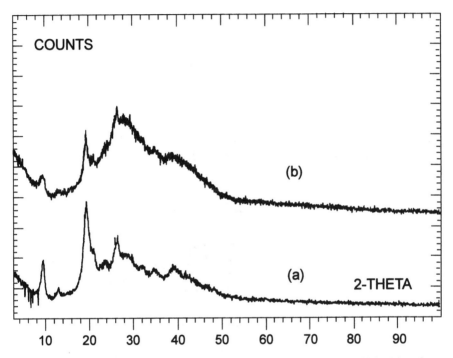

Figure 3. X-ray diffractograms of recovered chitin from crab α-chitin (a) and from squid β-chitin (b).

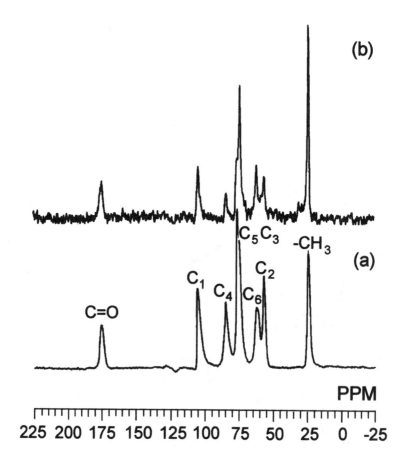

Figure 4. Solid state NMR spectra (CP-MAS ^{13}C NMR) of native squid β-chitin (a) and the corresponding recovered chitin (b).

of chitin into chitosan. As discussed in our previous publication (24), each swollen crystallite in the alkali complex acts as a separate reactor in the kinetics of deacetylation, a phenomenon referred to as " permutoid" reaction in the early cellulose literature (25). Hence, deacetylation of chitin only depends on the concentration of acetamide groups since alkali is in excess and the reaction is pseudo first order. The first step is swelling, which makes the chains of each microfibril uniformly accessible to alkali. Deacetylation, therefore, proceeds from within at temperatures above about 25°C. Once this pervading swelling takes place, each swollen crystallite is one phase saturated by alkali. The reaction is considered as homogeneous, although there is still a boundary between swollen crystallites and alkali solution. Therefore, the reaction rate depends only on the concentration of acetamide, but not on the diffusion of alkali ions. In this case, the plot of log concentration of acetamide group versus time is linear as shown in our previous paper (24). On the other hand, if there would be no swelling and complexation, the reaction, if present, would proceed from the microfibril surface inward. The implication of that would be a heterogeneous reaction leading to a mixture of chitosan and chitin, which has never been observed.

Conclusions

It is clear that antiparallel chain, polymorphic α-chitin undergoes permutoid swelling in alkali and recovers the same polymorphic crystal structure upon neutralization. It requires much higher concentrations of alkali than cellulose for mercerization, presumably because its unique amide intermolecular H-bonding has higher cohesive energy. Conversion of β-chitin to α-chitin through the alkali chitin complex is similar to the conversion of cellulose I to cellulose II (mercerized cellulose). In both cases, crystals containing a parallel packing of the chains are changed into an antiparallel packing without dissolution of the polymer. An intermediate alkali complex forms in which coalescence of adjacent swollen microfibrils having opposite polarity allows the chains to interdigitate and to adopt the antiparallel arrangement. Tentatively, the following schematic states the conclusions outlined above:

Of greater significance at this time is the role of alkali chitin in the transformation from α-chitin to chitosan, a valuable water soluble polyelectrolyte. As discussed in our previous publication (24), the kinetics of deacetylation of α-chitin follow a pseudo first order law; i.e., a log plot has linear dependence on the acetamide concentration. This was interpreted as due to each alkali chitin microfibril acting as a separate reactor with a finite acetamido concentration, but an excess of alkali in the reaction.

Acknowledgements

The authors thank Ms. Anne-Marie Lebuis and Dr. Fred Morin for obtaining the X-ray diffractograms and the solid state NMR spectra used in this paper.

Literature Cited

1. Robert, G. A. *Chitin Chemistry*; Macmillan Press Ltd.: London, **1992**.
2. Muzzarelli, R. A. A. *Chitin*; Pergamon Press: New York, NY, **1977**.
3. Neville, A.C. *The Biology of the Arthropod Cuticles*; Carolina Biological Readers, No 113, Burlington, Carolina Biology Supply Company, **1978**.
4. Neville, A. C. *Biology of Fibrous Composites: Development Beyond the Cell Membrane*; Cambridge University Press: New York, NY, **1993**.
5. Skjak-braek, G.; Anthonsen, T.; Sanford, P. *Chitin and Chitosan; Sources, Chemistry, Biochemistry, Physical Properties, and Applications, Proceedings from the 4th International Conference on Chitin and Chitosan*; Elsevier: New York, NY, **1989**.
6. Carlström, D. *J. Biophys. Biochem. Cytol.* **1957**, *3*, 669.
7. Tanner, S. F.; Chanzy, H.; Vincendon, M.; Roux, J. C.; Gaill, F. *Macromolecules*, **1990**, *23*, 3576.
8. Blackwell, J. *Biopolymers* **1969**, *7*, 281.
9. Rathke, T. D.; Hudson, S. M. *J. M. S.- Rev. Macromol. Chem. Phys.* **1994**, *C34(3)*, 375.
10. Hirano, S. In *Chitin Handbook*; Muzzarelli, R. A. A.; Peter, M. G., Eds.; Atec Edizioni, IT-63013 San Martino: Grottamare AP, Italy, **1997**; pp. 71-75.
11. Hirano, S. *Methods Enzymol.* **1988**, *161*, 408.
12. Thor, C. J. B.; Henderson, W. F. *American Dyestuff Reporter* **1940**, *29(19)*, 461.
13. Blackwell, J.; Kolpak, F.J.; Gardner, K.H. *TAPPI* **1978**, *61*, 71.
14. Revol, J.-F.; Goring D.A.I. *J. Appl. Polym. Sci.* **1982**, *26*, 1275.
15. Okano, T.; Sarko, A. *J. Appl. Polym. Sci.* **1984**, *29*, 4175.
16. Okano, T.; Sarko, A. *J. Appl. Polym. Sci.* **1985**, *30*, 325.
17. Kuang, S. J.; Revol, J.-F.; Goring, D. A. I. *Cellul. Chem. Technol.* **1985**, *19*, 113.
18. Revol, J.-F.; Dietrich, A.; Goring, D. A. I. *Can. J. Chem.* **1987**, *65*, 1724.
19. Bradford, H.; Revol, J.-F. In *Cellulose, Wood Chemistry and Technology, Proceedings of the 10th Cellulose Conference*; Schuerch, C., Ed.; Wiley Interscience, **1989**; pp. 129.
20. Li, J.; Revol, J.-F.; Marchessault, R. H. *J. Coll. Interface Sci.* **1996**, *183*, 365.
21. Focher, B.; Naggi, A.; Torri, G.; Cosani, A.; Terbojevich, M. *Carbohydrate Polym.* **1992**, *17*, 97.
22. Rudall, K. M. *J. Polymer Sci.* **1969**, *28*, 83-102.
23. Revol, J.-F.; Goring, D. A. I. *Polymer* **1983**, *24*, 1547.
24. Li, J.; Revol, J.-F.; Marchessault, R. H. *J. Appl. Polym. Sci.* **1997**, *65(2)*, 373.
25. Hermans, P.H. *Physics and Chemistry of Cellulose Fibres*; Elsevier Publishing Company Inc.: Amsterdam, **1949**.

Chapter 8

13C NMR Quantitative Techniques for Monitoring and Characterizing Biodegradable, Biocompatible Thermoplastics Produced by Bacteria

Using 13C NMR to Monitor and Characterize Poly(β-hydroxyalkanoates) in Living Cells and in Solution

Sheila Ewing Browne

Department of Chemistry, Mount Holyoke College, South Hadley, MA 01075

PHAs are produced by many types of bacteria as a carbon and energy storage material under conditions of nutrient limitation. As totally biodegradable, melt processable polymers, they are attracting considerable interest in applications for disposable plastic products, packaging materials and as biomedical polymers. PHAs are produced by industrial scale fermentation processes and sold by Monsanto under the trade name Biopol®. A new quantitative technique using natural abundance, solution 13C NMR was developed to monitor the intracellar formation and degradation in near real time of poly(β-hydroxyalkanoates)(PHAs) *in vivo* (in living cells) in *Alcaligenes eutrophus* and *Pseudomonas oleovorans*. The 13C NMR technique is a faster, easier and more accurate method than that of extracting the polymers from freeze dried cells. Because the measurement occurs in near real time, the NMR technique shows great potential for determining optimal harvest times during the fermentation process. It is more direct and sensitive than current monitoring methods (microscopy or optical density, OD). We have seen instances where using OD to determine harvest times would have resulted in a 30% lower yield of polymer than if 13C NMR were used. An additional 13C NMR technique was developed to determine the mole % of the subunits in PHA heteropolymers. It is a superior and more accurate technique for charaterization of PHAs than methanolysis of the isolated polymer and analysis by gas chromatography, saving both time and money.

Background on Biodegradable Polymers

Poly(β-hydroxyalkanoates)(PHAs), are accumulated by many diverse micro-organisms (over 600) as an intracellular energy and carbon storage material. They are produced by bacteria when they experience metabolic stress, such as a limitation of nitrogen, oxygen or other essential nutrients, while in the presence of excess carbon (*1-3*). PHAs, in essence bacterial fat, are ideal storage materials because they are in a highly reduced form and are water insoluble, inducing no osmotic pressure effects inside the cell.

PHAs are produced in natural environments by both aerobic and anaerobic bacteria from many different types of organic compounds, especially carbohydrates, organic acids, and hydrocarbons (4,5). In fact, these types of bacteria have been used in bioremediation of oil spills and chemical waste sites. As indicated by the generalized structure shown in Figure 1, naturally occurring PHAs, with very few exceptions, always contain repeating units with linear alkyl substituent groups at the β-position, with n = 0 - 8.

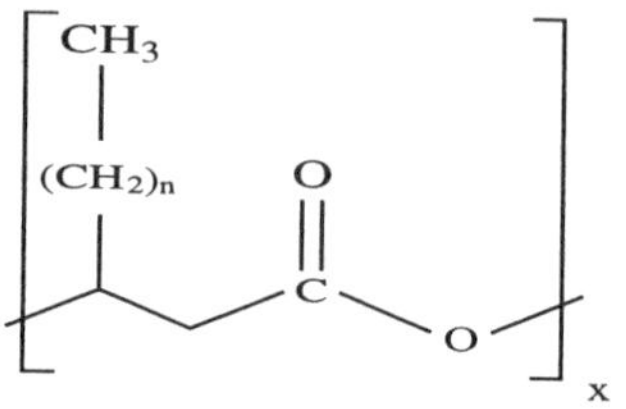

$$\left[\begin{array}{c} CH_3 \\ | \\ (CH_2)_n \quad O \\ | \quad \| \\ C \\ O \end{array} \right]_x \qquad n = 0 - 8$$

PHA

Figure 1. Structure of PHAs

The bacteria which produce PHAs can be divided into two groups, one which produces only short chain alkyl pendant groups, n = 0 and 1, and another which produces long alkyl chains, n = 2 or more. *Alcaligenes eutrophus* belongs to the first group and *Pseudomonas oleovorans* to the latter. It is very rare for bacteria to be able to produce both types of PHA. Each repeating unit of PHA has a chiral center at the β-position in the [R] configuration, giving perfectly isotactic polymers. This stereochemistry is essential to the inherent degradability of these biopolymers and is thought to be one of the main reasons for the slow rate or lack of biodegradability of chemically synthesized polyesters which have similar structures, but lack the stereoregularity of natural PHAs (6-8). PHAs are produced and stored inside granules in the bacterial cells, Figure 2. Two polymerases and a depolymerase, as well as structural proteins, are associated with the surface of the granules (9-11).

Poly(β-hydroxybutyrate) (PHB) was the first PHA to be discovered and was characterized as a high molecular weight polyester in 1925 by Lemoigne. It has received the most interest to date because it is the most widely naturally occurring PHA and it is completely biodegradable (to CO_2

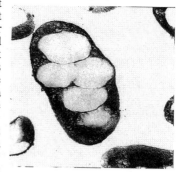

Figure 2. PHA in Granules

and H_2O) in fresh water, salt water, compost and sewer sludge. PHB, as a biocompatible polymer has great potential for use in biomedical applications (12,13) as a scaffold for cells in bridging deep wounds, in drug release systems, as artificial skin and in artificial tendons which again serve as a scaffold for cells to regrow. In these applications, the higher price of PHAs is not a deterrent. PHAs are degraded by two different mechanisms, hydrolysis and enzymatic catalysis. Hydrolytic degradation occurs in medical applications in humans, such as for the use of PHA as drug release carriers or surgical sutures. In environmental degradation, hydrolysis is too slow to be effective and instead bacteria and fungi produce extracellular enzymes which are capable of depolymerizing PHAs (14).

Figure 3. Degradation of PHB

PHB, however, has limitations for wide use as a thermoplastic. It forms a brittle plastic when molded due to its high degree and type of crystallinity (spherulitic). It is also difficult to melt process without thermal degradation. These problems were solved by inducing *A. eutrophus* to produce copolymers containing β–hydroxybutyrate and β-hydroxyvalerate, P(HB/V). Figure 3 shows spatulas made of P(HB/V) after 0, 4 and 6 months in a backyard compost pile in Massachusetts. Copolymers of poly(β–hydroxybutyrate-co-β–hydroxyvalerate, P(HB/V) are currently marketed by Monsanto as Biopol®. They are biodegradable thermoplastics which have lower degrees of crystallinity, smaller crystallite sizes, a lower melting temperature and much better mechanical properties than PHB (15).

P. oleovorans has been used to produce most of the long chain PHAs studied to date. Recent work has focused on the bacterial production of functionalized PHAs containing bromine, alkene, phenyl, nitrile, fluoride and ester moieties (14, 16-19). In general, *P. oleovorans* produces heteropolymers, such as PHO, which is a useful thermoplastic elastomer of low crystallinity, has a melting transition well below 100°C and a glass transition temperature well below 0°C. It is a rubbery material with a high molecular weight and good mechanical properties and is totally biodegradable to CO_2 and H_2O in the environment.

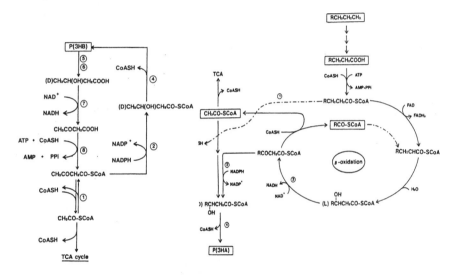

Figure 4. Biosynthetic Pathways for PHB (21,22) (left) and Long chain PHAs (20) (right)

The biosynthesis of PHAs vary according to the chain length produced, Figure 4 (20). PHB synthesis involves the formation of acetyl-coenzyme A (acetyl-CoA) from an appropriate carbon substrate, such as sugars, alcohols, or carbon dioxide, followed by the production of acetoacetyl-CoA and then 3-hydroxybutyryl-CoA. A sequence of three enzymatic reactions takes place in the formation of PHB. The three enzymes involved are 3-ketothiolase (acetyl-CoA acetyltransferase), acetoacetyl-CoA reductase (hydroxybutyryl-CoA dehydrogenase) and poly-β-hydroxybutyrate synthetase (steps 1-4 respectively on the left in Figure 4). In *P. oleovorans*, the pathway of PHA synthesis is linked to the cyclic β-oxidation and thiolytic cleavage of fatty acids (21). Heteropolymers are commonly produced in long chain PHAs, n = 2 - 8, due to the condensation of acetyl-CoA with an acyl-CoA (resulting in the addition or removal of one molecule of acetic acid) in the biosynthetic step catalyzed by 3-ketothiolase, Figure 5 (21). If bacteria are fed an 8 carbon substrate, the resulting polymer will contain subunits of 8, 6 and 10 carbons. For example, when *P. oleovorans* was fed n-octanoic acid, it produced poly(β-hydroxyoctanoate), PHO, which contained subunits of n = 2 (10.7%), n = 4 (83.0%) and n = 6 (6.0%) (22,23).

Monitoring PHAs Quantitatively Using a New ^{13}C NMR Technique

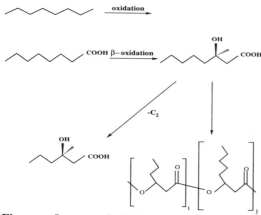

Figure 5. β-Oxidation and Thiolytic Cleavage of Fatty Acids

We have developed at Mount Holyoke College a quantitative ^{13}C NMR technique to determine the rates of formation and degradation of PHAs *in vivo*. The concentration of both short and long chain polymers and the presence of functionalized side chains are determined in near real time. A slurry of bacterial cells containing PHAs is prepared and the NMR spectra are obtained at set elevated temperatures following our standard parameters for measurement.

In the past, quantitatively monitoring the PHAs produced by bacteria was done by centrifuging the culture, lyophilizing the cells and extracting the polymer into refluxing CH_2Cl_2 for 24 hours (*14*). After the extraction, the cellular debris was removed by filtration and the remaining PHA, which was in solution, was precipitated in methanol, collected by filtration and weighed. To determine the ratio of subunits present, the PHA obtained by the extraction was then methanolyzed in refluxing methanol and acid to form the β-hydroxy methyl esters. These were evaluated by gas chromatography with the use of standards which had to be synthesized.

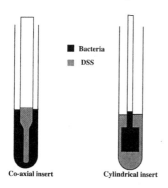

Co-axial insert   Cylindrical insert

■ Bacteria
▨ DSS

Figure 6. NMR Samples

The concentration of PHA can now be obtained directly from the living cells without the extensive sample preparation when using the quantitative ^{13}C NMR technique that we have developed. To obtain an NMR spectrum, an aliquot of the cell culture is spun down and the cells are resuspended as a slurry in D_2O in a 10mm NMR tube. We have used DSS, sodium 2,2-dimethyl-2-silapentane-5-sulfonate, $[(CH_3)_2SiCH_2CH_2SO_3Na]$, as a standard in either a coaxial insert or in a cylindrical insert (for very small samples) (Figure 6). This is necessary because the bacteria can metabolize the standard.

The first ^{13}C NMR spectrum of a PHA *in vivo* was obtained by Saunders for PHB (*24*). We have developed a quantitative method using ^{13}C NMR spectroscopy and have obtained for the first time the spectra of many PHAs, including poly(β-hydroxybutyrate)(PHB), poly(β-hydroxyoctanoate)(PHO), poly(β-hydroxynonanoate)(PHN), poly(β-hydroxyphenylvalerate), PHPV, a blend of both PHN and PHPV and funtionalized PHAs containing alkenes, bromine, epoxides and esters *in vivo* in *A. eutrophus* and *P. oleovorans*. The ^{13}C NMR spectra of PHAs *in vivo* show only the carbons of the PHAs in the bacterial cells. The many other carbon containing compounds have peaks which are not resolved from the noise.

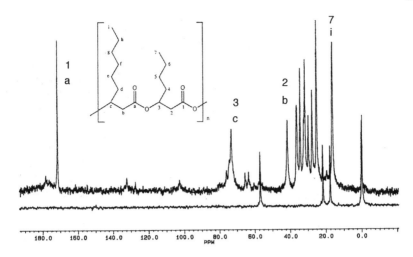

Figure 7. ^{13}C spectra of PHN *in vivo* in *P. oleovorans* with DSS (top) and of DSS alone (bottom)

Figure 7 shows the structure of PHN, with its two most abundant subunits, and the spectra of PHN *in vivo* in *P. oleovorans* with DSS in a coaxial insert. A spectrum of DSS with four peaks is shown for reference below the PHN spectra. The carbonyl peak of PHN at 172 ppm is relatively sharp. The peaks for the other two backbone carbons, the methine (CH) peak at 74 ppm and the methylene (CH$_2$) peak at 47 ppm, as well as the peaks due to the alkyl carbons in the β alkyl side chain (between 15 and 40 ppm), are broader. The pulse widths, delay times, temperature, line broadening and number of scans must be adjusted to obtain a reasonable signal to noise for each PHA. For the PHN spectrum in Figure 7, these were 6µs, 2s, 314K, LB = 10 Hz and 2,000 scans, respectively. Sample size has varied from 100 mL of culture to a liter depending on the density of cells and concentration of PHA. We were able to determine the presence of PHA formation before it could be seen by microscopy with or without Nile Blue. The time to obtain a spectrum varied from 15 minutes for PHB or PHO samples to two or more hours for PHPV, which was produced by *P. oleovorans* in very low yields (*25*). We have followed the formation or degradation of several PHAs *in vivo* in both *P. oleovorans* and *A. eutrophus* by taking the spectrum of cells taken at various time intervals from the culture of bacteria after it had been fed a carbon source. For degradation studies of PHPV, PHN and a blend of both, the bacteria were allowed to produce the polymers and were then spun down and resuspended in a culture medium containing no carbon source. The degradation of the PHA by the bacteria was followed by taking aliquots of this culture of cells at intervals of approximately 20 hours. Changes in the ratio of peak height of the carbonyl of the PHA (at 172 ppm) relative to the peak height of the DSS standard (at 0 ppm), which is constant over time, allows changes in concentration of the PHA to be measured.

When *P. oleovorans* is fed nonanoic acid or 5-phenylvaleric acid or a mixture of both, the ^{13}C NMR spectra clearly show PHN (middle), PHPV(top) and the blend of both (bottom) *in vivo* (Figure 8). The ^ symbol denotes peaks due to DSS and the numbers and letters refer to the structure in Figure 7. The peak at 172 ppm for the carbonyl of PHN was used to follow PHN degradation and the peak due to the ortho, meta and para carbons of the phenyl group of PHPV at 130 ppm was used to follow PHPV degradation. The series of spectra of the blend of PHA and PHPV, shown in Figure 9, established that the "unnatural" PHPV polymer was biodegradable by the

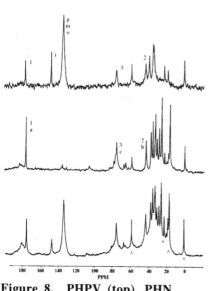

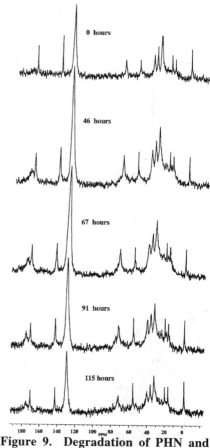

Figure 8. PHPV (top), PHN (middle) and PHN, PHPV blend (bottom)

Figure 9. Degradation of PHN and PHPV *in vivo* at 318K

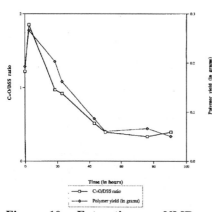

Figure 10. Extraction vs. NMR

bacteria producing it, although the rate of degradation was much slower that that of PHN. This blend of PHPV and PHN, one a homopolymer and one a heteropolymer, were present in the same bacterial cell and in the same granules (26).

By correlating the grams of PHN obtained by the extraction method (mentioned earlier) with the ratio of peak heights, (C=O/DSS), one can then calculate the concentration of PHA in other samples using only the NMR spectra. Figure 10 shows the excellent correlation between the extraction method and the ^{13}C NMR method for following the degradation of PHN.

Spectra taken *in vivo* at several different temperatures for PHB, PHO and PHPV can be seen in Figures 11, 12 and 13, respectively. The mobility of the polymers can be evaluated by examining the peak broadening at lower temperatures. A broader peak is indicative of slower motion. The backbone carbons (47 and 74 ppm) are the broadest at all temperatures and, therefore, the least mobile in all three polymers. In fact, the backbone carbons almost disappear at room temperature. The sharpness of the backbone carbonyl is presumably due to less scalar broadening from protons. The pendant alkyl groups (17 to 40 ppm) are more mobile than the backbone carbons and are still visible, although quite broad, at room temperature. Looking at the spectra at the lowest temperature for all three polymers, it is clear that PHB has the sharpest alkyl peaks. The mobility of PHB is greater that PHO and both are much more mobile than PHPV *in vivo*. This difference in mobility made it possible to use the carbonyl peak at 318K to follow PHN degradation in a blend of PHN and PHPV because, at this

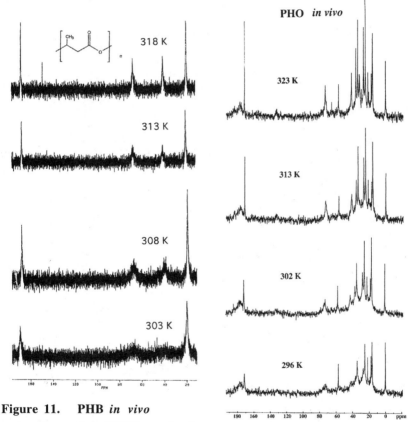

Figure 11.   PHB *in vivo*

Figure 12.   PHO *in vivo*

temperature, the PHPV peak has broadened into the base line. The fact that we can see peaks of PHAs in the NMR spectra proves that all of these polymers *in vivo* are quite amorphous, but have very different degrees of mobility. Crystalline PHAs would have peaks too broad to show up in solution NMR. The restricted mobility of PHPV *in vivo* may explain its slower rate of formation and degradation. It may also be the cause of the poor yield of this polymer. We were able to use this information on mobility to examine the mobility of PHA in granules isolated for study of depolymerase activity. The mobility of the isolated granules was the same as that of granules *in vivo*, showing that the process of isolation had not caused crystallization of the polymer.

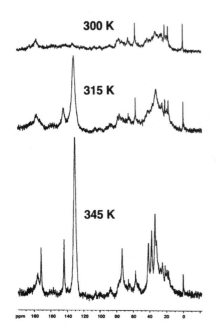

300 K

315 K

345 K

ppm 180 160 140 120 100 80 60 40 20 0

Figure 13.   PHPV *in vivo*

Characterization of PHAs in Solution

^{1}H NMR spectroscopy has been used for characterization of short chain PHAs, such as P(HB/V), after extraction from bacteria. ^{1}H NMR spectroscopy is certainly a less time consuming method of analysis than the method used for long chain PHAs which involves extraction, methanolysis and analysis by gas chromatography. Futhermore, it does not require standards. Figure 14 shows the ^{1}H and ^{13}C NMR spectra of a copolymer of P(HB/V) in DCCl$_3$. The ratio of the areas of peak 3 to peak 4 for the CH$_3$ group in the ^{1}H spectrum gives the ratio of HB to HV in the copolymer. Unfortunately, ^{1}H NMR spectroscopy can not be used for the determination of the quantity or type of PHAs *in vivo* due to the presence of water (H$_2$O). The large peak due to H$_2$O *in vivo* in ^{1}H NMR obscures the peaks of the hydrogens in the polymer. ^{1}H NMR also has little applicability for the long chain polymers involved in our research, due to peak overlap for the alkyl hydrogens of the

polymer. The ^{13}C NMR spectrum of P(HB/V) in Figure 14 shows all the different carbons, with the exception of the carbonyl carbons, clearly resolved.

Using   natural   abundance solution ^{13}C NMR to characterize polymers has several advantages. Most   NMR   characterization   of polymers is done with solid state NMR, a more expensive, complex and less generally available instrument than that used for solutions.   Solid or solution ^{13}C NMR spectra have much higher resolution than those of ^{1}H NMR. Solution ^{13}C NMR spectra can be obtained in both aqueous and hydrocarbon solvents.

We have developed a new technique using ^{13}C NMR, to determine the % ratio of subunits in P(HB/V) and are now applying the technique to long chain PHAs.   In most applications using ^{13}C NMR spectra, the size of a peak for a particular carbon in the spectra has no correlation with the number of carbons of that type in the compound.   The

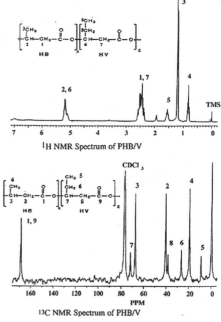

^{1}H NMR Spectrum of PHB/V

^{13}C NMR Spectrum of PHB/V

Figure 14.   P(HB/V) in DCCl$_3$

peak size is more indicative of the number of protons attached to that type of carbon. Carbonyl carbon peaks are smaller than those of carbons bonded to hydrogen. This is due to the method of obtaining ^{13}C NMR spectra which requires that the hydrogens be decoupled from the carbons. A Nuclear Overhauser Effect, NOE, causes carbons attached to hydrogens to have an enhanced signal. By using delay times which allow all of the NOE to transfer to the carbon and then comparing the peak areas of carbons with the same number of hydrogens in the different subunits, the ratios of the subunits can be obtained.

Table I Composition of the P(HB/V) in mole % HV using ^1H and ^{13}C NMR.

Percentage of HV reported by Aldrich	Percentage of HV from ^1H NMR	Mean percentage HV from ^1H NMR	Percentage of HV from ^{13}C NMR	Mean percentage HV from ^{13}C NMR
	11.78		11.67	
11	12.28	12	10.38	11
	12.08		11.76	
	26.50		24.54	
21	27.46	26	23.51	23
	27.00		22.32	
	52.83		50.44	
46	53.43	53	51.29	51
	53.38		50.70	

Figure 14 (bottom) shows the ^{13}C NMR spectra of the same sample of P(HB/V) as in the ^1H NMR spectra. The ratio of peak area for peak 4 and 5 for the methyl groups in HB and HV gives the ratio of HB to HV in the copolymer. Table 1 shows a good correlation between the mole % determined by ^{13}C NMR and ^1H NMR. The samples of P(HB/V) were purchased from Aldrich and the percentage of HV which they reported did not give the uncertainty of their analysis (done using ^1H NMR).

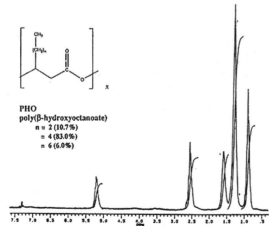

PHO
poly(β-hydroxyoctanoate)
n = 2 (10.7%)
  = 4 (83.0%)
  = 6 (6.0%)

Figure 15. ^1H NMR of PHO

The ^1H NMR spectrum of PHO in Figure 15 clearly shows that ^1H NMR cannot be used to evaluate the ratio of subunits for PHO. The peaks have too much overlap to distinguish between the C6, C8 and C10 subunits.

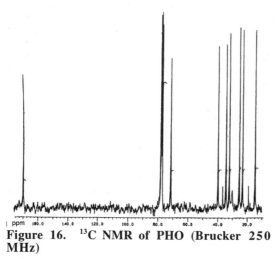

Figure 16. ¹³C NMR of PHO (Brucker 250 MHz)

The ¹³C NMR spectra of PHO taken on Bruker 250 and 500 MHz NMR spectrometers, Figures 16 and 17, show much better resolution of the peaks than in the ¹H NMR spectrum. We are currently assigning the alkyl peaks in PHO and PHN spectra. It has been necessary to synthesize methyl β-hydroxyalkanoates as standards to insure the correct assignment. When peaks due to carbons in different subunits are assigned, ¹³C NMR can be used to determine the ratio and mole % of the subunits for these long chain polymers in $DCCl_3$. We hope to extend this characterization to polymers *in vivo*. It is clear that the broader peaks *in vivo* compared to *in vitro* (in $CDCl_3$) (Figures 7, 16 and 17) may present a problem for PHO and PHN, but will certainly work for P(HB/V). Use of a higher field in the cases of PHO and PHN *in vivo* may help to increase the resolution.

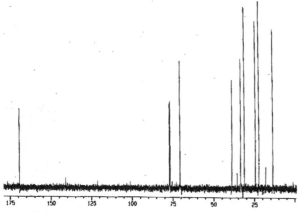

Figure 17. ¹³C NMR of PHO (Bruker 500 MHz NMR)

Conclusions

We have developed at Mount Holyoke College a quantitative ¹³C NMR technique to determine the rates of formation and degradation of PHAs *in vivo*. The concentrations of both short and long chain polymers and the presence of functionalized side chains were determined in near real time. It was successfully used to measure the composition and concentration of PHAs in living cells. This technique can be applied to obtain rapid feedback on conditions necessary to optimize yield and composition of desired PHAs in bacteria. This will allow the tailoring of natural biodegradable polymers for specific applications. The use of natural abundance solution ¹³C NMR for characterizing subunits of copolymeric PHAs was proven to be accurate. These new techniques are superior to currently used quantification and characterization methods for PHAs.

Acknowledgments

The author gratefully acknowledge the National Science Foundation, Grant No. MCB-9202419, the General Electric Foundation "Faculty for the Future" Grant to the Mount

Holyoke College and Chemistry Department of Mount Holyoke College. The following Mount Holyoke undergraduate students were involved in the research described in this paper: Chelvanaya Gabriel, Anastasia Dimitropoulou, Pamela Maynard and Sapana Panday. R. W. Lenz, R. C. Fuller, E. S. Stuart and M. Rothermich of the University of Massachusetts, Departments of Polymer Science and Engineering (Lenz) and Biochemistry and Molecular Biology, were collaborators in this research, supplying bacterial samples for NMR analyses. I would like to thank Leta Bouse and Rodolfo Akel for help in preparing the manuscript.

References

1. Langeveen, R. G.; Huisman, G. W.; Preusting, H.; Ketelaar, P.; Eggink, G.; Witholt, B. *Applied and Environmental Microbiology* **1988**, *54*, 2924-2932.
2. Gross, R. A.; DeMello, C.; Lenz R.W.; Brandl, H.; Fuller R.C. *Macromolecules* **1989**, *22*, 1106-1115.
3. Kim, Y. B. ; Kim, Y. B., Ed.; Dept. of Polymer Science & Engineering, University of Massachusetts, Amherst, Ma., 1989.
4. Doi, Y.; Segawa, A.; Kawaguchi, Y.; Kunioka, M. *FEMS Microbiology Letters* **1990**, *67*, 165-170.
5. Anderson, A. J.; Dawes, E. A. *Microbiological Reviews* **1990**, *54*, 450-472.
6. Hocking, P. J.; Timmins, M. R.; Scherer, T. M.; Fuller, R. C.; Lenz, R. W.; Marchessault, R. H. *Journal of Macromolecular Science-Pure and Appl. Chem.* **1995**, *A32*, 889-894.
7. Hocking, P. J.; Marchessault, R. H.; Timmins, M. R.; Lenz, R. W.; Fuller, R. C. *Macromolecules* **1996**, *29*, 2472-2478.
8. Fritzsche, K., Lenz, R. W.; Fuller, R. C. *Int. J. Biol. Macromol.* **1990**, *12*, 92.
9. Fuller, R. C.; O'Donnell, J. P.; Saulnier, J.; Redlinger, T. E.; Foster, J.; Lenz, R. W. *FEMS Microbiol. Rev.* **1992**, *103*, 279.
10. Stuart, E. S.; Lenz, R. W.; Fuller, R. C. *Can. J. Microbiol.* **1995**, *41*, 84-93.
11. Stuart, E. S.; Foster, L. J. R.; Lenz, R. W.; Fuller, R. C. *Biological Macromolecules* **1996**, *19*, 171-176.
12. Gross, R. A. *Bacterial Polyesters*; Hanser Publishers: New York, 1994.
13. Dawes ; Kluwer Academic Publishers: Nordrecht, 1990.
14. Kim, Y. B.; Lenz, R. W.; Fuller, R. C. *Macromolecules* **1991**, *24*, 5256-5260.
15. Lenz, R. W. **1995**, *45*, 15-20.
16. Kim, Y. B.; Lenz, R. W.; Fuller, R. C. *Macromolecules* **1992**, *25*, 1852-1857.
17. Lenz, R. W.; Kim, B. W.; Ulmer, H. W.; Fritzsche, K.; Knee, E.; Fuller, R. C. *Functionalized Poly-B-Hydroxyalkanoates Produced by Bacteria*; Lenz, R. W.; Kim, B. W.; Ulmer, H. W.; Fritzsche, K.; Knee, E.; Fuller, R. C., Ed.; Kluwer Academic Publishers: Boston, 1990; Vol. 186, pp 23-35.
18. Scholz, C.; Fuller, R. C.; Lenz, R. W. *Makromol. Chem.* **1993**, *194*.
19. Lenz, R. W.; Ulmer, H.; Kim, Y. B.; Fuller, R. C.; Plenum Press: New York, 1992; Vol. 7.
20. Peoples, O. P.; Sinskey, A. J. *Polyhdroxybutyrate (PHB): A Model System for Biopolymer Engineering: II*; Kluwer Academic Publishers: Nordrecht, 1990.
21. Doi, Y. *Microbial Polyesters*; VCH Publishers: New York, 1990.
22. Gagnon, K. D.; Fuller, R. C.; Lenz, R. W.; Farris, R. J. *Rubber World* **1992**, 32.
23. Brandl, H.; Gross, R. A.; Lenz, R. W.; Fuller, R. C. *Applied and Environmental Microbiology* **1988**, *54*, 1977-1982.
24. Amor, S. R.; Rayment, T.; Sanders, J. K. M. *Macromolecules* **1991**, *24*, 4583-4588.
25. Curley, J. M.; Lenz, R. W.; Fuller, R. C.; Browne, S. E.; Gabriel, C. B.; Panday, S. *Polymer* **1997**, *38*, 5313-5319.
26. Curley, J. M.; Lenz, R. W.; Fuller, R. C. *International Journal of Biological Macromolecules* **1996**, *19*, 29-34.

POLYMER SYNTHESIS AND DEGRADATION

Chapter 9

Enzymology of the Synthesis and Degradation of Polyhydroxyalkanoates

Geoffrey A. R. Nobes and Robert H. Marchessault

Department of Chemistry, McGill University, 3420 University Street, Montreal, Quebec H3A 2A7, Canada

Polyhydroxyalkanoates are an important class of biodegradable polymers. This review discusses the enzymology of the biosynthesis and biodegradation of these materials. A brief history of PHA research, from their discovery in bacteria earlier this century to the present, is provided, as is a discussion of the physical properties of PHAs, including comparisons to other common polymers. Research on the biosynthesis and pathways involved in PHA production is reviewed with a focus on the key enzymes involved in these processes. One of the main strategic advantages of PHAs is their inherent biodegradability. This review also covers the biodegradation of PHAs, both intracellularly and extracellularly. Again, the focus is primarily on the depolymerase enzymes which microorganisms efficiently use for the exploitation of PHAs as an energy source.

Polyhydroxyalkanoates (PHAs) are a class of thermoplastic polyesters produced by a number of species of bacteria. The biosynthesis of intracellular PHAs is part of the β-oxidation cycle and is prompted by conditions of stress, such as the shortage of an essential nutrient like phosphate or nitrogen (*1*). PHAs occur as intracellular inclusions (or granules after they have been isolated from the bacterial cell). The most common member of the PHA family is poly(3-hydroxybutyrate), P(3HB), where R in Figure 1 is a methyl group. P(3HB) is produced as a polymer of 10^5 to 10^6 molecular weight in inclusions of 0.2 to 0.5 μm in diameter (*2*). Each inclusion contains approximately 1000 polymer chains. P(3HB) is a form of carbon that is osmotically inert (*3*). P(3HB) inclusions contain approximately 97 to 98% P(3HB), 2% protein and 0.5% lipid. The lipid is a mixture of phosphatidic acid, triacetin, tributyrin, tripropionin and other, unidentified lipids (*4*).

Short chain length (SCL)

R = H, C H$_3$, C$_2$H$_5$
x = 1 - 3
thermoplastics

Medium chain length (MCL)

R = C$_3$H$_7$ - C$_{13}$H$_{27}$
x = 1 - 4
elastomers

Figure 1. General structural formulae of bacterial polyhydroxyalkanoates.

PHAs have been suggested as a fifth class of physiologically important biopolymers, along with polyisoprenoids, polypeptides, polysaccharides and polynucleotides (5). P(3HB) granules *in vivo* are noncrystalline, but, once isolated, are found to be 60 to 70 % crystalline. This implies the influence and interaction of water within the bacterial cell (5) and Lauzier *et al.* have proposed a model with specific hydrogen-bonding of nascent P(3HB) with water (6).

There are two main classes of PHAs: short chain length (SCL) with 3 to 5 carbon atoms per repeat unit and medium chain length (MCL) with 6 to 14 carbon atoms in the repeat unit (7, 8). The changeover from SCL to MCL coincides with a change in physical properties from crystalline thermoplastic to low crystallinity elastomeric thermoplastic (1).

PHAs degrade to water and carbon dioxide in the environment and are especially of interest because of this biodegradability. Several applications of PHAs have been developed in the past decade which exploit their inherent biodegradability (9). One of the most common applications is bottles. In 1990, Wella, a German hair care company, released shampoo bottles made from Biopol (a copolymer of P(3HB-co-3HV). Biopol bottles have also been available in the USA as of 1991 and in Japan as of 1992 (9). A bicycle helmet was made of reinforced Biopol fiber with knitted cellulosic high performance fibers to optimize strength and energy absorbance behavior. The helmet was completely degraded in the soil after 40 days (9). A biodegradable autoseparative filter was made from 90% NaCl and 10% Biopol. After thermal consolidation, salt crystals were dissolved away and an open porous filter matrix remained (9).

Historical Perspective. P(3HB) was discovered in *Bacillus megaterium* by Maurice Lemoigne of the Institut Pasteur, Paris, in 1925 (10, 11). Lemoigne also reported finding P(3HB) in plants as well as in human blood (12, 13). P(3HB) was the only PHA known until 1964, when Davis reported the biosynthesis of poly(3-hydroxybutyrate-co-3-hydroxy-2-butenoate) (7). In 1974, Wallen and Rohwedder isolated a heteropolymer in activated sludge containing repeat units with up to seven carbon atoms and the presence of two 3-hydroxyacids in polymers isolated from activated sewage sludge (14, 15). Nine years later, White and coworkers reported that *B. megaterium* accumulated a polymer of 95% 3HB, 3% 3-hydroxyheptanoate and 2% of an eight carbon 3-hydroxyacid with trace amounts of three other 3-hydroxyacids, including 3-hydroxypropionate. Around the same time, de Smet *et al.* reported the occurrence of granules of poly(3-hydroxyoctanoate) in *P. oleovorans* (7). Now over 91 different repeat units are known to occur in PHAs (1) and over 90 genera of bacteria, both Gram-negative and Gram-positive, have been found that accumulate PHAs (8). The majority of PHAs are poly R-(-)-3-hydroxyalkanoates containing between 3 and 14 carbon atoms per repeat unit.

The 91 different hydroxyalkanoic acids known to occur in PHAs reveals the low substrate specificities of PHA synthases (16). The known monomers of the PHA family include almost the complete range of 3-hydroxyalkanoic acids, from 3-hydroxypropionic acid to 3-hydroxyhexadecanoic acid. As well, there are unsaturated 3-hydroxyalkenoic acids with one or two double bonds in the side chain, 3-hydroxyalkanoic acids with methyl groups at various positions in the side chain, non 3-hydroxyalkanoic acids, such as 4-hydroxybutyric acid, 4-hydroxyvaleric acid, 4-hydroxyhexanoic acid, 4-hydroxyheptanoic acid, 4-hydroxyoctanoic acid, 4-hydroxydecanoic acid, 5-hydroxyvaleric acid, 5-hydroxyhexanoic acid, and 6-hydroxydodecanoic acid. Also found are 3-hydroxyalkanoic acids with various functional groups in the side chain, such as free carboxyl groups (as found in malic acid), alkyl ester, benzoic acid, acetoxy, phenoxy, phenyl, or cyclohexyl groups. In addition, there may be halogen atoms in the side chain or double bonds in the backbone, such as in 3-hydroxy-2-butenoic acid or 6-hydroxy-3-dodecenoic acid or 3-hydroxyalkanoic acids with a methyl at the α-carbon (e.g., lactic acid). All of these acids form hydroxyalkanoate-CoA thioesters and are substrates for PHA synthesis *in vivo* (*16*).

A bacterial copolymer of poly(3-hydroxybutyrate-*co*-3-hydroxyvalerate) was commercialized by Imperial Chemical Industries (and later Zeneca), under the tradename Biopol, starting in the early 1980s at approximately 320 000 to 500 000 kg/yr using *Alcaligenes eutrophus* (*1, 2, 7*). The cost was approximately $15 per kg, in contrast to about $1 per kg for commodity plastics. There are two main sources of the cost, the carbon source (glucose or fructose) and the isolation of the polymer. The use of alternative carbon sources, such as cellulose, hemicellulose, molasses and corn syrup is being explored. Genetic engineering also holds promise, as it may be possible to alter the substrate specificity of the enzymes involved in PHA biosynthesis.

The possibility of using plants as a PHA production source is being investigated. *Arabidopsis thaliana* is a small weed in the mustard family and makes granules like in bacteria, which under laboratory conditions, has produced PHA in up to 15 to 20 % yield with the cloned genes for acetoacetyl-CoA reductase and PHA synthase (*A. thaliana* has its own β-ketothiolase) from *A. eutrophus* (*1, 7*). The PHA is produced in the cytoplasm, nucleus and vacuoles in individual cells and in the leaf, cotyledon and root tissues of the plants. There is a desire to place PHA genes into the seeds of plants, such as canola, where both the valuable oil and PHA would be harvested. Producing PHA in plants may reduce the cost of its production (*2*). This is theoretically possible because of the high level of metabolic flux through acetyl-CoA, the substrate for P(3HB) biosynthesis.

Recently, P(3HB) production has been reported in transgenic cotton (*17*). *A. eutrophus* genes encoding acetoacetyl-CoA reductase (*phaB*) and PHA synthase (*phaC*) were put into cotton cells by particle bombardment. (β-ketothiolase activity is present in cotton.) Fibers of ten transgenic plants expressed *phaB* and eight expressed *phaB* and *phaC*. These plants produced P(3HB) granules with an average polymeric molecular weight of 0.6×10^6 to 1.8×10^6 in the cytoplasm. The presence of P(3HB) in the cotton fibers yielded changes in the thermal properties (*17*). The fibers exhibited better insulating characteristics. The rate of heat uptake and cooling were slower, which meant a higher heat capacity.

PHAs often exist as copolymers. Hence, complex carbon sources can produce PHAs with interesting compositions (*16*). Only the *R*(−) configuration is detected at the chiral center for biosynthetic PHA. Long chain PHAs (i.e., hydroxyalkanoic acids with more than 16 carbons) have not yet been found.

Very few bacteria make copolymers of 3-hydroxybutyrate (or other SCL-PHAs) with MCL-PHA (*16*). An exception is *Rhodococcus ruber*, which can accumulate a polymer containing 3-hydroxybutyrate, 3-hydroxyvalerate, and 3-hydroxyhexanoate when grown on hexanoic acid (*18*). Most PHAs contain two or more hydroxyalkanoic acids. Some PHAs, such as P(3-hydroxyvalerate), P(3-hydroxyhexanoate), P(3-hydroxyoctanoate) and P(4HB), are rarely found as homopolymers. P(3HV) homopolymer has been found in *Chromobacterium violaceum* in amounts up to 65 % dry cell weight (*7, 19*). P(3HV) has also been found in *Chromatium purpuratum*, *Paracoccus denitrificans* and *Rhodospirillum rubrum* (*7, 20*).

The genes for the enzymes involved in P(3HB) biosynthesis from *A. eutrophus* have been cloned and transferred into *E. coli*, which can then produce P(3HB) *in vitro*. PHB-producing *E. coli* has also been equipped with a gene to lyse the cells at a selected time to release the P(3HB) granules into the medium (*7*). The *in vitro* production of P(3HB) has been commercialized by Metabolix Inc. of Cambridge, MA, whose process produces a polymer of up to 1.3×10^7 molecular weight by a pseudoemulsion polymerization (*1, 21*).

PHB with a 1-4 linkage have also been found, which eliminates the asymmetry present in PHAs with 1-3 linkages (*1*). Polymers with 1-4 linkages are more mobile (*22*) than those with 1-3 linkages, which are typically rigid with higher melting and glass transition temperatures. P(4HB) homopolymer has been produced in *Alcaligenes eutrophus* H16 (*7*). The incorporation of different units into the PHA chain is possible as a result of the broad substrate specificity of the enzymes involved in PHA biosynthesis (*8*).

Terminology and Nomenclature of PHAs. The terminology suggested by Steinbüchel (*23*) will be used in this review. PHA is a generic term. P(3HB) is used if analysis has been done to confirm the homopolymer. P(3HB-*co*-3HV) refers to a copolymer, while P(3HB)/P(3HO) refers to a blend. SCL refers to PHAs with between 3 and 5 carbon atoms per repeat. MCL is used for PHAs with between 6 and 14 carbon atoms per repeat. LCL is used for PHAs with more than 14 carbon atoms per repeat.

The enzymes for PHA biosynthesis should be termed PHA synthases, as opposed to PHB synthase, since no single PHA synthase has been demonstrated to use only one particular hydroxyacyl-CoA thioester substrate (*23*). *phaC* is the general term for the PHA synthase structural gene. Biosynthesis genes are named *phaA*, *phaB*, *phaC*, etc., while those coding for PHA degradation are named *phaZ*, *phaY*, *phaX*, etc. There is typically only SCL- or MCL-PHA in a given bacterium, not both at significant molar fractions (*23*).

Physical Properties of PHAs

Generally, the physical properties of P(3HB) are similar to those of isotactic polypropylene (24). A comparison of some physical properties of P(3HB) and polypropylene is provided in Table I. P(3HB) has a melting point similar to polypropylene, better O_2 barrier properties than polyethylene terephthalate and polypropylene, mechanical properties similar to polystyrene and polypropylene, and a water vapor transmission rate three times lower than polypropylene (25). A comparison of P(3HB) with other common biodegradable polymers is given in Table II and the chemical formula are shown in Figure 2.

Table I. Physical Properties of P(3HB) and Polypropylene[a]

Property	P(3HB)	Polypropylene
Molecular Weight	5.0×10^5	2.0×10^5
Crystalline Melting Point (°C)	175	176
Crystallinity (%)	80	70
Glass Transition Temperature (°C)	4	-10
Density (g/cm^3)	1.25	0.905
Young's Modulus (GPa)	4.0	1.7
Tensile Strength (MPa)	40	38
Elongation to Break (%)	6	400

[a]Source: Adapted from ref. 26.

P(3HB) is a noncrystalline, mobile elastomer *in vivo*, but rapidly crystallizes upon isolation (27). Water presumably plays a major role in plasticizing noncrystalline P(3HB) within the bacterial cell (6) or in preventing crystallization. Also, it is thought that polymer mobility is prerequisite for susceptibility to intracellular depolymerase.

All P(3HA)s have unit cells of the $P2_12_12_1$ space group with the lattice parameters varying depending on the length of the side chain and the comonomer composition. The P(3HB) unit cell was first reported by Okamura and Marchessault (29) and subsequently refined by several groups (1, 30-32). The unit cell is orthorhombic with space group $P2_12_12_1$-D_2^4 and dimensions $a = 0.576$ nm, $b = 1.320$ nm and c (fiber repeat) $= 0.596$ nm. The unit cell contains two left-handed helical molecules packed antiparallel.

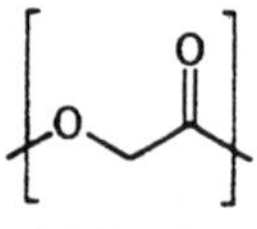

poly(glycolic acid)

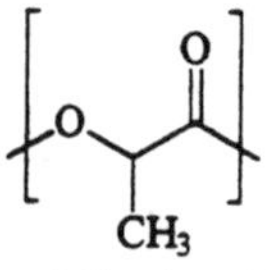

poly(lactic acid)

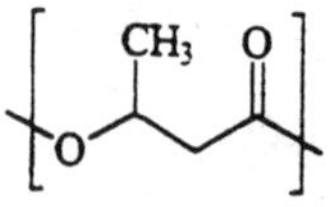

poly(3-hydroxybutyrate)

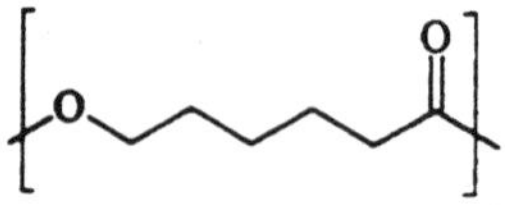

poly(ε-caprolactone)

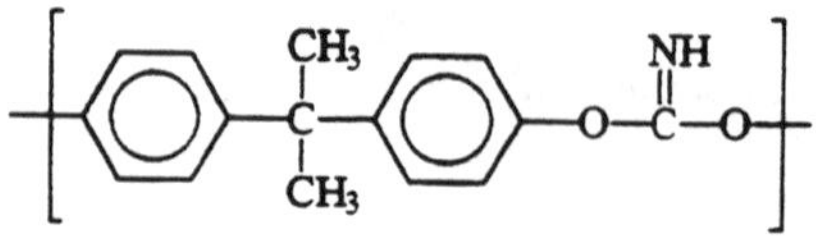

poly(Bisphenol A iminocarbonate)

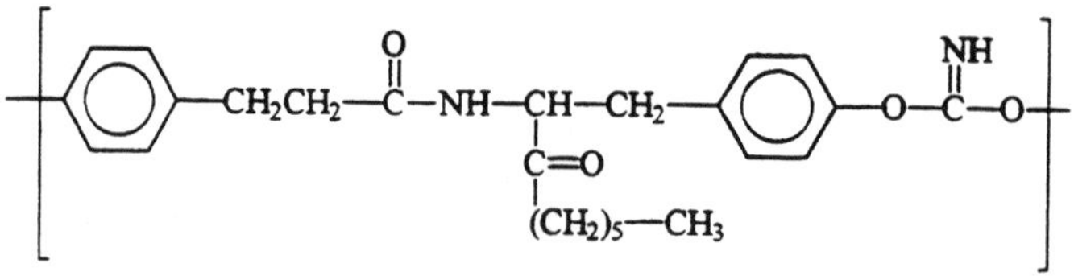

poly(DTH iminocarbonate)

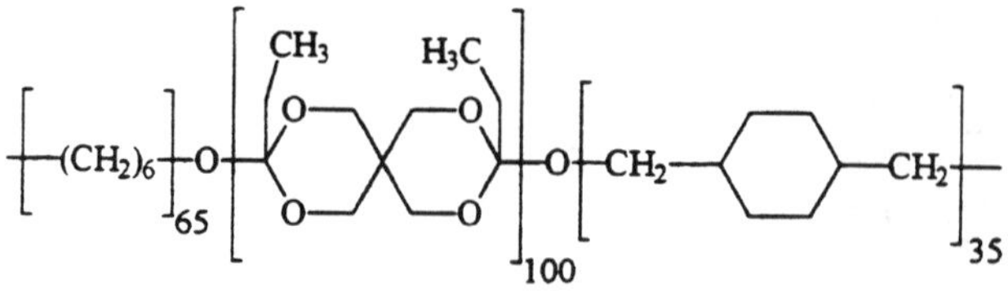

poly(DETOSU-HD-CDM ortho ester)

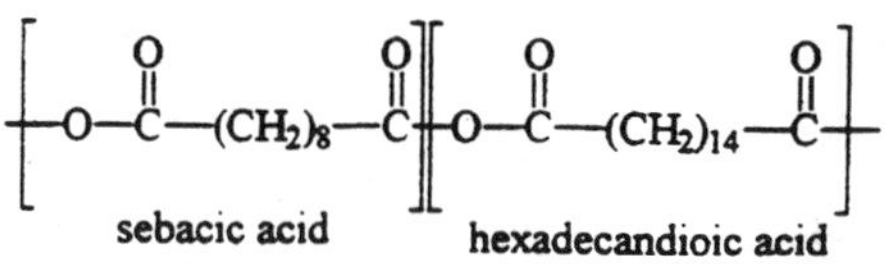

sebacic acid   hexadecandioic acid

poly(SA-HDDA anhydride)

Figure 2. Structural formulae of biodegradable polymers listed in Table II.

Table II. Mechanical Properties of Common Degradable Polymers[a]

Polymer	T_g (°C)	T_m (°C)	Tensile Strength (MPa)	Young's Modulus (MPa)	Elongation to Break (%)
Poly(glycolic acid) (50,000 mw)	35	233	na	na	na
Poly(L-lactic acid) (50,000 mw)	54	170	28	1200	6.0
Poly(L-lactic acid) (300,000 mw)	59	178	48	3000	2.0
Poly(D,L-lactic acid) (107,000 mw)	51	-	29	1900	6.0
Poly(D,L-lactic acid) (550,000 mw)	53	-	35	2400	5.0
Poly(3-hydroxybutyrate) (422,000 mw)	1	180	36	2500	2.5
Poly(ε-caprolactone) (44,000 mw)	-62	57	16	400	80
Poly(SA-HDDA anhydride) (142,000 mw)	na	49	4	45	85
Poly(DETOSU-HD-CDM ortho ester) (99,700 mw)	55	-	20	820	220
Poly(Bisphenol A iminocarbonate) (105,000 mw)	69	-	50	2150	4.0
Poly(DTH iminocarbonate) (103,000 mw)	55	-	40	1630	7.0

[a]The structures of the polymers listed in this table are provided in Figure 2.
Source: Adapted from ref. 28.

The P(3HV) unit cell was reported by Yokouchi *et al.* (*31*) for racemic synthetic polymer and is also an orthorhombic unit cell ($P2_12_12_1$-D_2^4) with dimensions $a = 0.932$ nm, $b = 1.002$ nm and c (fiber repeat) $= 0.566$ nm. The unit cell contains two left-handed helical molecules packed in an antiparallel fashion. Subsequently, bacterial P(3HV) was shown to crystallize in the same unit cell and space group (*33*).

In copolymers of P(3HB-*co*-3HV), the phenomenon of isodimorphism occurs where cocrystallization of 3HB and 3HV repeat units takes place (*24, 34*). As the content of HV increases, the (110) plane of the P(3HB) lattice expands and only the a parameter of the unit cell changes. A eutectic point is found in the plot of melting temperature versus composition (see Figure 3) at 30 to 40 mol% HV, where the P(3HB) crystal lattice switches to the P(3HV) lattice.

The crystallization of MCL-PHAs is significantly different from that of SCL-PHAs (*24*). When there are 3 to 4 carbons in the side chain, the polymers do not crystallize. When there are 5 to 7 carbons in the side chain, the polymers have low crystallinity (*35, 36*). These polymers crystallize in a 2_1 helix with two molecules per orthorhombic unit cell. The fiber repeat (c dimension of the unit cell) approaches 0.45 nm with increasing side chain length. This highlights the packing effects of the side chain, whose mutual interactions cause the backbone helix to compress. Long side chains form ordered sheets with extended *n*-alkane branches (*35, 36*). Side chains with 3 to 4 carbons cannot crystallize and the main chain cannot crystallize either. Therefore, these PHAs remain in a noncrystalline state.

The melting temperature (T_m) of P(3HB) is 179°C and that of P(3HV) is 112°C (*24*). The glass transition temperature (T_g) of P(3HB) is approximately 0°C. P(3HB) is brittle, but P(3HB-*co*-3HV) becomes rubbery with increasing HV content. Figure 4 shows the effect of HV content on Young's modulus. At room temperature, the Young's modulus of P(3HB) is 3.8 GPa, while for P(HB-*co*-28%HV) it is 1.5 GPa. The highest elongation of P(3HB) is 65% at 100°C, while for P(HB-*co*-28%HV) it is 750% at 90°C. At 25°C, the tensile strength of P(3HB) is 44 MPa, which decreases with increasing 3HV up to 20%. Elongation to break increases from 3 to 27% as 3HV goes from 0 to 20 mol% (*24*). At 28 mol% HV, elongation to break is 700%. This increase in the toughness of P(3HB-*co*-3HV) copolymers compared to the P(3HB) homopolymer is remarkable because the crystallinities are comparable due to isodimorphism.

For P(3HA)s with long side chains, the elongation to break is approximately 300% and the Young's modulus is typical of thermoplastic elastomers, that is, about a tenth of the value for thermoplastic P(3HB-*co*-3HV) copolymers (*36*). Figure 5 shows the change in glass transition temperature as a function of the length of the side chain. Figure 6 shows stress-strain curves and Young's modulus for SCL-PHAs and MCL-PHAs.

Formation of PHA

PHAs are formed intracellularly as inclusions (or granules). Aspects of PHA biosynthesis are similar to emulsion polymerization as was first suggested by Ellar

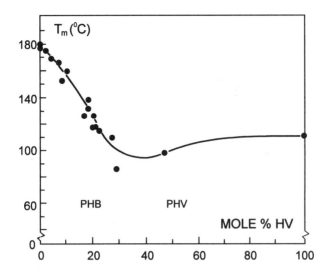

Figure 3. Melting point versus composition curve for bacterial P(3HB-*co*-3HV). Between 30 and 40 mol % of HV, the crystal lattice changes from that of P(3HB) to that of P(3HV). (Reproduced with permission from ref. 37. Copyright 1988 Verlag.)

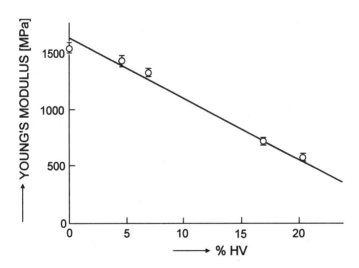

Figure 4. Effect of HV content on Young's modulus for P(3HB-co-3HV) bacterial copolymers. (Reproduced with permission from ref. 37. Copyright 1988 Verlag.)

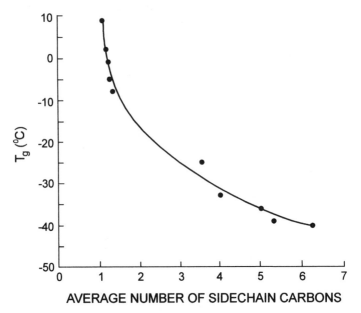

Figure 5. Effect of increasing the average number of carbons in the side chain of PHAs on the glass transition temperature (T_g). (Reproduced with permission from ref. 36. Copyright 1990 Butterworth & Co.)

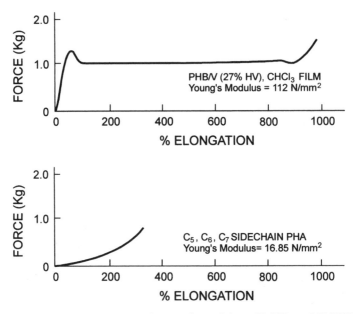

Figure 6. Stress-strain curves and Young's modulus of P(HB-*co*-27%HV) and MCL-PHAs. (Reproduced with permission from ref. 36. Copyright 1990 Butterworth & Co.)

et al. in 1968 (*38*). Individual granules are regarded as latex particles (*39*). Granule formation may be caused by the formation of a micelle by the polymerase molecules. Another idea is that granule formation is like a homogeneous nucleation mechanism (*40*), where the polymerase molecules which propagate in the cytoplasm form precursor granules. The precursor granules then grow until they are mature granules.

Emulsion polymerization is a heterogeneous reaction system, just as P(3HB) polymerization is inherently heterogeneous (*39*). For granule formation, the monomer reacts with the polymerase in the cytoplasm to produce an oligomer-synthase complex which is surface active (the polymerase is hydrophilic, the P(3HB) tail is hydrophobic). One or several growing chains precipitate to form a colloidal particle, the micelle. It is proposed that this attracts phospholipids and other cellular compounds for stabilization, as in stage I of an emulsion polymerization.

During stage II, there is a constant number of granules per cell (for *A. eutrophus* 8 to 14) and there is a constant rate of polymerization (*39*). The monomer diffuses to the granule where it is polymerized at the surface by the polymerase. Also, it has been suggested that once a given polymerase chain transfers, it will desorb from the granule surface due to its hydrophilic character. Monomer is supplied by a cascade of reactions in the cytoplasm.

Under conditions of nutrient limitation (such as nitrogen or phosphorous), PHA synthesis acts as a carbon reserve and electron sink for the reducing power which accumulates as a consequence of oxygen limitation (*8, 15*). Electrons are no longer able to cross the electron transfer chain to oxygen at the same rate and are diverted to the reductive step of polymer synthesis. Under oxygen limitation, the $NADH/NAD^+$ ratio rises, the TCA cycle slows down, PHA synthesis essentially captures some of the excess reducing power, reduces the $NADH/NAD^+$ balance, and relieves the inhibition of the TCA cycle This enables the cycle to operate faster than would otherwise have been possible and, when growth limitation is alleviated, P(3HB) is catabolized to acetyl-CoA.

During PHA production, most of the carbon source is used for PHA, but some must be oxidized for energy generation (*7*). Energy is required to drive the PHA biosynthetic pathway (for example, NADPH is consumed by the reductase) and for cellular maintenance reactions and the cellular energy level must be maintained so that the overall pattern of regulation favors PHA production. If a supplemental energy source is used, then a larger fraction of the carbon source is able to be converted into related monomers and, thus, increase PHA yield (*7*). The use of an alternate energy source may also modulate the cellular energy level.

Biosynthesis of PHAs. A plethora of reports on the biosynthesis and growth of PHAs in many different organisms, using many different substrates, exists in the literature. The examples cited in this section of the review were chosen to highlight certain aspects of the substrate specificities of the enzymes involved in PHA biosynthesis.

The biosynthesis of PHAs can be summarized as follows (*16*). First, a carbon source enters the organism by diffusion or by specific transport through the cytoplasmic membrane. Then, anabolic or catabolic reactions (or both) convert the compound into hydroxyacyl-CoA thioesters. Then, the PHA synthase polymerizes the hydroxyacyl-CoA thioesters to form PHAs. This second step can occur in several ways; e.g., some bacteria convert acetyl-CoA to acetoacetyl-CoA to 3-hydroxybutyryl-CoA which then forms P(3HB). A schematic of the biosynthesis of PHAs is shown in Figure 7.

Most pseudomonads belonging to the ribosomal rRNA homology group I, which includes *P. oleovorans* and *P. putida*, accumulate PHA copolymers with repeat units containing between 6 and 14 carbons when the cells are grown on alkanoic acids (*7, 8*). The polymer which is produced is related to the substrate used for growth. For example, *P. oleovorans* grown on octanoate accumulates a PHA random copolymer of 89% 3-hydroxyoctanoate and 11% 3-hydroxyhexanoate, whereas growth on dodecanoate generates 31% 3-hydroxydodecanoate, 36% 3-hydroxydecanoate, 31% 3-hydroxyoctanoate and 2% 3-hydroxyhexanoate.

Brandl *et al.* grew *P. oleovorans* on all *n*-alkanoic acids from formate to decanoate (*41*). The best growth results were obtained with octanoate and nonanoate. For those with 4 carbons or more, *P. oleovorans* always formed the acetate which was subsequently used to produce the polymer. (The acetate is a metabolic intermediate.) This suggests that during cell growth, 2-carbon fragments are cleaved from the initial carbon substrate by β-oxidation, which would provide energy for the bacterial cell via the tricarboxylic acid (TCA) cycle. Although *P. oleovorans* grew on the *n*-alkanoic acids, intracellular PHAs were only detected for acids with 6 or more carbons. The major constituent of the polymer always had the same number of carbons as the *n*-alkanoic acid used for growth. On octanoate, nonanoate and decanoate, a large fraction of the repeat units contained two carbons fewer than the *n*-alkanoic acid. For hexanoate, the polymer contained 18% 3-hydroxyoctanoate and 82% 3-hydroxyhexanoate. When heptanoate and octanoate were used, again the polymer contained significant amounts of repeat units which were 2 carbons longer or shorter than the substrate. This suggests that *P. oleovorans* can produce acetate from the *n*-alkanoic acid by β-oxidation and then use this for building longer chain length compounds for incorporation into the PHA (*41*).

For *n*-alkanoic acids with odd numbers of carbon, small amounts of even chain lengths were found. For even chain lengths, small amounts of odd chain lengths were found (*41*). This suggests a decarboxylation mechanism by which CO_2 is produced from the growth substrate to form intermediates which are one carbon shorter. These intermediates can also be incorporated into the PHA. For those with one carbon longer, partial decarboxylation and subsequent incorporation of the acetate into the PHA is expected.

Since the degree of polymerization for PHAs from caproate and heptanoate is approximately twice that for octanoate, nonanoate and decanoate, Gross and coworkers suggest a mechanism of polymerization (*35*). An overall relatively higher efficiency of enzymatic conversion from a carbon source to polymer does not mean that the final enzymatic step of the polymerization shows a similar specificity for the predominant monomer structure. The degree of polymerization is dependent on the binding constant (K_m) of the polymerase to the chain end. K_m may be a function of

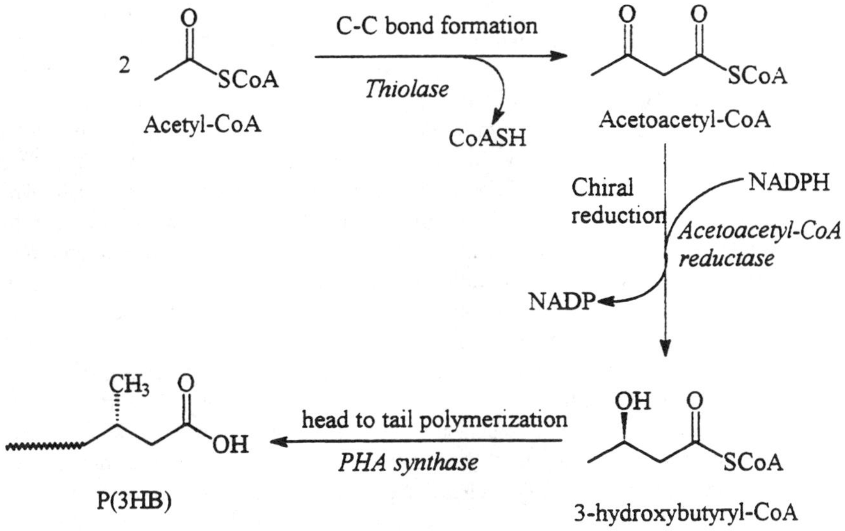

Figure 7. The general three-enzyme biosynthetic pathway leading to PHAs.

both the substrate-enzyme fit as well as the hydrophobicity of the terminal repeat unit of the growing chain which is bound to the enzyme. A pendant group with a longer chain length would introduce greater hydrophobicity to the active site of the polymerase.

Although in *A. eutrophus* it was suggested that P(3HB) metabolism allows for simultaneous synthesis and degradation (*42*), the regulation of metabolism is not understood in detail. Prior to complete utilization of the carbon substrate in the media, the cells decrease the concentration of polymerase while increasing the depolymerase. This appears to represent the cell's response to an increasingly less permissive metabolic environment (*43*).

The relationship between fatty acid metabolism and PHA biosynthesis in *P. putida* was examined by Eggink *et al.* (*44*). During growth on glucose, the 3-hydroxyacyl-acyl carrier protein (ACP) intermediates of the *de novo* fatty acid biosynthetic pathway are diverted to PHA biosynthesis. Figure 8 gives a summary of major pathways involved in PHA biosynthesis, including alkanoate elongation, β-oxidation and fatty acid biosynthesis. Also, during cultivation on fatty acids, intermediates of the β-oxidation cycle serve as precursors of PHA biosynthesis. The fatty acid biosynthetic pathway in bacteria is responsible for the synthesis of saturated and unsaturated acyl moieties of membrane lipids. The ratio of saturated to unsaturated is temperature dependent in order to maintain proper cellular membrane flexibility. The degree of unsaturation of PHA and membrane lipids is similarly influenced by shifts in the cultivation temperature. *P. putida* at 15°C had twice the amount of unsaturated monomers as found in PHA accumulated at 30°C. It is not clear if (*R*)-3-hydroxyacyl-ACP is a substrate for the PHA polymerizing system or if the transfer of an acyl moiety to CoA is required prior to polymerization (*44*).

Long chain fatty acids can also be used as substrates for PHA biosynthesis (*44*). In this case, fatty acids are degraded by β-oxidation until medium chain length intermediates are formed. This implies that the length of the PHA produced depends on the PHA synthesizing system, the chemical structure of the long chain fatty acid and the fatty acid degradation pathway. Degradation of fatty acids requires the involvement of additional enzymes (i.e., more than just for β-oxidation). For example, enoyl-CoA isomerase is involved in the β-oxidation of long chain fatty acids with double bonds extending from odd-numbered carbon atoms (*44*). It is not known if β-oxidation and *de novo* fatty acid biosynthesis can operate simultaneously to generate PHA precursors (if so, it may be possible to prepare PHAs from sugars and vegetable oils).

Fukui *et al.* (*46*) studied P(3HB) synthesis in *Zoogloea ramigera*. *Z. ramigera* is a floc forming microorganism which accumulates P(3HB) rapidly with simultaneous flocculation. The P(3HB) synthase is present not only in the granules, but also in the soluble fractions. The intracellular distribution of the enzyme activity seems to depend on the presence of carbohydrates in the medium. The organism grows dispersedly in the absence of glucose and forms only a small amount of P(3HB), with the synthase activity mainly in the soluble fraction. With the addition of glucose, the bacterium starts to flocculate and accumulate P(3HB) granules

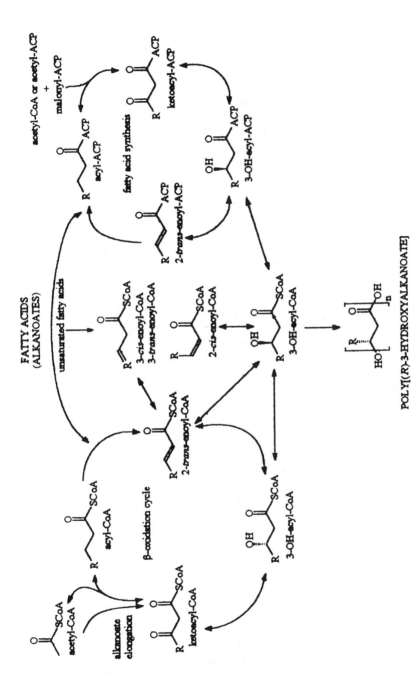

Figure 8. The major pathways involved in PHA biosynthesis are alkanoate elongation, β-oxidation and fatty acid biosynthesis. (Adapted from ref. 45.)

rapidly. Enzyme activity in the soluble fraction becomes associated with the newly formed granules (46).

The biosynthesis of P(3HB-co-3HV) in R. ruber was reported by Anderson and coworkers (47). The production of the 3HV monomer from glucose in R. ruber confirmed the carboxylation of pyruvate, followed by the TCA cycle reaction to yield succinate, which is the precursor of propionyl-CoA which leads to 3HV. Direct carboxylation of pyruvate may occur, followed by reverse reactions of the TCA cycle to yield succinyl-CoA. It was estimated that this route would account for approximately 12% of the carbon flux in the synthesis of 3HV (47).

Although the majority of bacteria studied posses one of two types of PHA synthase, which allows for the synthesis of either SCL-PHAs or MCL-PHAs, a few are able to accumulate both types (8). P. aeruginosa grown on valerate produces a PHA copolymer composed of 3HV to 3-hydroxydecanoate and P. resinovorans grown on hexanoate accumulated 3HB-3-hydroxyhexanoate-3-hydroxyoctanoate-3-hydroxydecanoate. Also, the phototropic bacteria Rhodospirillum rubrum grown on hexanoate and Rhodocyclus gelatinous grown on heptanoate gave rise to PHAs containing 3HB, 3HV and 3-hydroxyhexanoate repeat units and 3HB, 3HV and 3-hydroxyheptanoate copolymers, respectively (8).

Pathways of PHA Biosynthesis. Four general pathways of PHA biosynthesis have been identified (7). Representative organisms for these pathways are *Alcaligenes eutrophus, Rhodospirillum rubrum, Pseudomonas oleovorans* and *Pseudomonas aeruginosa*. In each of these organisms, polymerization is catalyzed by a PHA synthase (also referred to as PHA polymerase) and the substrate for PHA biosynthesis is usually a hydroxyacyl-coenzyme A (hydroxyacyl-CoA) (7). There are two classes of PHA synthases; those for short chain length (SCL) hydroxyacyl-CoAs (3-5 carbons) and those for medium chain length (MCL) hydroxyacyl-CoAs (6-14 carbons). Although hydroxyacyl-CoAs are synthesized in many metabolic pathways, a distinct pathway leading to PHAs exists (7).

The PHA biosynthetic pathway uses acetyl-CoA, which generates 3-hydroxybutyryl-CoA precursors, and may use propionyl-CoA to generate 3-hydroxyvalerate-CoA (7). There are usually three enzymes in the PHA biosynthesis; β-ketothiolase, acetoacteyl-CoA reductase and PHA synthase. In some organisms, especially those producing MCL-PHAs, β-ketothiolase and the reductase are not present. Two forms of both β-ketothiolase and acetoacetyl-CoA reductase have been found in *A. eutrophus* (7). The two ketothiolases have different substrate specificities, whereas the reductases differ in their specificity for nicotinamide dinucleotide factors. The reductase for PHA synthesis is specific for NADPH and the other is specific for NADH.

In *Pseudomonas aeruginosa* and *Pseudomonas putida*, the monomer precursors are apparently produced in the fatty acid biosynthetic pathway (7). These organisms produce mainly 3-hydroxyoctanoate and 3-hydroxydecanoate from simple compounds, such as glucose and glycerol.

The condensation of acetyl-CoA (C_2 unit) to form acetoacetyl-CoA (C_4 unit) is a bio-Claisen reaction (48). In all bio-Claisen reactions, the anion derived from acetyl-CoA acts as a nucleophile which reacts with the ketone or thiol ester. Each biosynthetic thiolase has a different role in ketone body utilization, generating acetoacetyl-CoA as a substrate for steroid and isoprenoid biosynthesis or initiating a three enzyme pathway leading to P(3HB).

The role of β-oxidation of short chain alkanoates in PHA copolymer synthesis was examined in *Azotobacter vinelandii* UWD (*49*). β-oxidation likely directs valerate into PHA synthesis. The short chain fatty acids butyrate, valerate, *trans*-2-pentenoate, crotonate, hexanoate, heptanoate and octanoate induced the production of the β-oxidation enzymes enoyl-CoA hydratase and *L*-(+)-3-hydroxybutyryl-CoA dehydrogenase. 3-hydroxyacyl-CoA epimerase activity in the β-oxidation complex was not detected. 3-ketoacyl-CoA, the product of *L*-(+)-3-hydroxybutyryl-CoA dehydrogenase, was directed into PHA synthesis through acetoacetyl-CoA reductase which produced the 3HV repeat unit (*49*). If valerate was used as the sole carbon source, most of the valerate was directed into metabolism and very little PHA was formed. When glucose was also present, the β-oxidation of short chain fatty alkanoates inhibited the activity of acetoacetyl-CoA reductase and 3-ketothiolase and, thus, the PHA yield. This suggests that the use of fatty acids to promote PHA copolyester formation in *A. vinelandii* will result in decreased PHA yield (*49*).

Enzymology of PHA Biosynthesis. The understanding of the enzymology of PHAs is very important for the improvement of production methods. In this section, the first two enzymes in the three enzyme PHA pathway will be briefly considered, while PHA synthases will be considered in more detail in the next.

In *A. eutrophus*, there are two 3-ketothiolases which together accept 3-ketoacyl-CoAs with between 4 and 10 carbons (*8, 15*). In addition, the acetoacetyl-CoA reductase is active with 3-ketoacyl-CoAs of 4 to 6 carbons to generate *R*-(-)-3-hydroxyacyl-CoAs. The PHA synthase polymerizes 3-hydroxy, 4-hydroxy and 5-hydroxyalkanoates from 3 to 5 carbons, but not 6 carbons or longer. There are also two ketothiolases in *Zoogloea ramigera*. There are also two isozymes of acetoacetyl-CoA reductase, one which is NADH dependent and another which is NADPH dependent. Acetoacetyl-CoA reductase is five times more active with NADPH than with NADH.

The regulation of PHA synthesis occurs by inhibition of the β-ketothiolase by free CoA (*7, 15*). The reversible reaction, catalyzed by 3-ketothiolase, is inhibited in the condensation direction by free CoA. The ratio of NADH to NAD^+ also affects PHA synthesis. During cell growth, the CoA level in the cells is high due to the rapid flux of acetyl-CoA into the TCA cycle. When a nutrient, such as the nitrogen source is exhausted, the $NADH/NAD^+$ ratio increases, inhibiting the enzymes of the TCA cycle. As the flux of acetyl-CoA decreases, CoA levels decrease, removing the inhibition of β-ketothiolase. Acetyl-CoA or propionyl-CoA may enter the PHA biosynthetic pathway to produce 3HB or 3HV monomers. Many PHA monomer precursors, hydroxy-CoAs from other biosynthetic pathways, do not require the action of β-ketothiolase or acetoacetyl-CoA reductase. Inhibition by CoA does not, however, affect the incorporation of these monomers into PHAs during cell growth (7).

The *Zooglea ramigera* thiolase is coded by 1176 nucleotides, corresponding to a 392 residue subunit of M_r = 40.8 kDa (48). The native enzyme is a homotetramer of 163 kDa. This thiolase was used as a hybridization probe for the *A. eutrophus* thiolase. 248 out of 392 residues were identical for 63% absolute homology.

The thiolase treats two identical molecules of acetyl-CoA differently (48). The first is used at carbon-2 as a nucleophile and the other at carbon-1 as an electrophile (i.e., to produce a head-to-tail condensation). The thermodynamics of thiolase reactions strongly favor thiolytic cleavage of acetoacetyl-CoA. The K_{eq} value has been estimated to be between 1.5×10^{-5} and 6×10^{-6}. Despite the unfavorable thermodynamics, acetoacetyl-CoA is drawn off biosynthetically with reductase for P(3HB) accumulation.

The proposed active site of the thiolase includes 3 cysteines and 1 histidine (48). In the first half reaction, a cysteine attacks acetyl-CoA to form an acetyl-thio-enzyme intermediate, which then reacts in the second half reaction with the anion to form the second acetyl-CoA by enzymatic deprotonation.

Over 25 genes involved in PHA biosynthesis in various bacteria have been cloned (23). These genes include the 3-ketothiolase (*phaA*), acetoacetyl-CoA reductase (*phaB*), PHA synthase (*phaC*) and PHA depolymerase (*phaZ*). In *A. eutrophus*, *phaA*, *phaB* and *phaC* are part of a single operon (see Figure 9). Expression of these three genes in *E. coli* enables it to synthesize up to 80 % dry weight of P(3HB). Inclusions are of similar size and appearance to *A. eutrophus* inclusions.

PHA Synthases. The key enzyme in PHA biosynthesis is the PHA synthase. Steinbüchel *et al.* have provided molecular data for PHA synthases from *A. eutrophus*, purple nonsulfur bacteria (such as *R. rubrum*), purple sulfur bacteria (such as *Chromatium vinosum*), pseudomonads belonging to rRNA sequence homology group 1 (such as *Pseudomonas aeruginosa*), *Methylobacterium extorquens* and Gram-positive bacteria (such as *Rhodococcus ruber*) (23). The nucleotide sequences of 10 different PHA synthase genes from 8 different bacteria have been obtained (23). Three different types of PHA synthases can be distinguished with respect to their substrate specificity and structure.

Type I comprises synthases of bacteria which synthesize SCL-PHAs, including *A. eutrophus* and *R. ruber*. These synthases exhibit 37 to 39% identical amino acids. Type II includes the PHA synthases of *P. oleovorans*, *P. aeruginosa* and *P. putida*, which synthesize MCL-PHAs. These synthases exhibit 54 to 80% identical amino acids. The level of similarity between type I and type II synthases is between 34 to 40% identical amino acids. Type III PHA synthases are found in the bacteria *Chromatium vinosum* and *Thiocapsa violaceae*. These bacteria synthesize SCL-PHAs. Their *phaC* genes encode proteins 35 to 40% smaller than type I or type II synthases and show weak homology to them, with only 21 to 27% identical amino

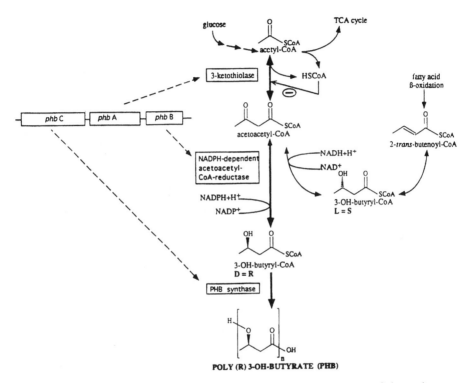

Figure 9. The organization of the genes encoding the enzymes of the major pathway leading to P(3HB) in *Alcaligenes eutrophus*. The *phaA* gene codes for the thiolase, which is inhibited by CoA. Two acetyl-CoA molecules are joined by the thiolase to form acetoacetyl-CoA. The *phaB* gene codes for the acetoacetyl-CoA reductase, which requires NADPH. The reductase forms 3-hydroxybutyryl-CoA, which is the substrate for PHA synthase, the product of the *phaC* gene. (Reproduced with permission from ref. 45. Copyright 1995 National Research Council of Canada.)

130

acids. It is thought that, in *C. vinosum* and *T. violaceae*, expression of a second gene (*phaE*), located near *phaC*, may be required for PHA synthase activity. This raises the possibility that two activities, encoded by distinct polypeptides, may be required for PHA synthase activity in these bacteria. All PHA synthases show some homology (*23*). Little is known about the tertiary and quaternary structures of PHA synthases or their reaction mechanisms (*50*).

The genes encoding the enzymes for the P(3HB) pathway in *A. eutrophus* were cloned in the late 1980s. Since then, PHA synthases from an additional 17 different bacteria have been cloned (*23*). The organization of the biosynthetic genes is as follows. The genes are often clustered in the bacterial genomes. In *A. eutrophus* H16, the structural genes for the PHA synthase (*phaC*), 3-ketothiolase (*phaA*) and acetoacetyl-CoA reductase (*phaB*) are organized in that order in a single operon. In other bacteria, PHA synthases are not clustered with genes coding for other proteins required for PHA synthesis. In *Z. ramigera*, *phaA* and *phaB* make up a single operon with the *phaC* gene not adjacent. *P. oleovorans* and *P. aeruginosa* each possess two genes coding for PHA synthase which are separated by a gene encoding a PHA depolymerase.

Detailed study of PHA synthase and its reaction is hampered by difficulties in purifying the enzyme to homogeneity (*8*). Early experiments showed that PHA synthase is bound to inclusions, whereas the synthase in soluble fractions of bacteria was not active. Recently, the PHA synthase of *A. eutrophus* was cloned and expressed in *E. coli*. *In vitro* mutagenesis of the gene, followed by purification and an assay of the synthetic activity, have shown that only one of two cysteine residues that are conserved between several synthases was essential for enzyme activity. Also, in *E. coli*, the PHA synthase was found to be posttranslationally modified by the addition of 4'-phosphopantetheine. These data support a model of the PHA synthase reaction which resembles that proposed for fatty acid synthase. In the model, the growing PHA chain cycles between a phosphopantetheine and a cysteine thiol to permit the alternate loading and elongation steps involved in PHA synthesis (*8*).

The P(3HB) synthase from *A. eutrophus* has been well studied (*23, 25, 51-53*). The *A. eutrophus* PHA synthase exists as an equilibrium mixture of monomeric and dimeric forms. The rate of polymer formation experiences a long and variable lag which is dependent on the protein concentration. The lag may be the result of the assembly of the dimeric form needed for activity and/or a priming process that may involve CoA esters of oligomeric 3-hydroxybutyrate (*53*). Gerngross and Martin also suggest that granule assembly may be responsible for the lag (*21*). Similar lags are observed for other enzymes in biosynthetic polymerizations, such as glycogen synthase and rubber polymerase (*53*).

The residue cysteine-319 is the only conserved cysteine in all PHA synthases examined to date and site-directed mutagenesis has shown that it is essential for catalysis (*51, 53*). Acylation of this residue causes a shift of the monomeric form of the synthase to its oligomeric form, which leads to a substantial increase in its specific activity and a substantial decrease in the lag of polymer formation. How this relates to the *in vivo* priming of the synthase is not known.

Liebergesell *et al.* (*50*) studied the PHA synthase from *C. vinosum* and its location at the granule surface. The PHA synthase of *C. vinosum* is a type III synthase. A two-step mechanism was proposed involving thiol groups and the formation of an acyl-S intermediate, which is then transferred to the polyester primer (*54*). The expression of PHA synthase in *C. vinosum* requires the expression of both the *phaC* and *phaE* genes. This may mean that there are two subunits of the synthase since the PhaC protein is 39.5 kDa and the PhaE protein is 40.5 kDa. The active synthase is very complex, being composed of ten subunits (*50*). After cloning and expression of the synthase, it was possible to produce P(3HB) *in vitro*, but only in 10 to 15 % yield. The PHA synthase was bound to the granule surface.

Both 3-ketothiolase and acetoacetyl-CoA reductase are soluble, cytosolic enzymes, whereas PHA synthase is attached to the P(3HB) granule (*52*). The role of PHA synthase in the accumulation of granules was discussed by Gerngross *et al.* (*52*). There are two main issues. The first is the mechanism by which the synthase initiates and carries out the condensation of thousands of monomers to produce the polymer. The second is the role of the synthase in granule assembly. Immunocytochemical localization placed the PHA synthase on the surface of the granule. In whole cells of *A. eutrophus* in the absence of a carbon source, the synthase is essentially randomly distributed throughout the cytoplasm. If a carbon source is added, granules accumulate and the synthase is found at the surface of the granules. It is not known if the synthase is at the surface because of covalent linkages to the growing polymer chain or simply a hydrophobic interaction (*52*).

Gerngross *et al.* proposed a model for granule assembly as follows (*52*). In the first step, the metabolism of the cell reaches a level where free CoA becomes depleted and acetyl-CoA accumulates and *R*-3-hydroxybutyryl-CoA is produced. Secondly, the synthase is primed by an unknown mechanism, polymerization takes place and polymer chains grow by extrusion from the synthase. Then the increased chain length makes the growing polymer chain more hydrophobic, which leads to the assembly of amphipathic molecules in the aqueous cytoplasm which, in turn, leads to micelle formation. This is followed by continued polymerization and granule growth. As the polymer continues to accumulate, the granules are forced to fuse together. The driving force for the reaction is the availability of substrate and free energy generated by the release of free CoA. In the model, no membrane or lipids are required for granule assembly (*52*).

Gerngross *et al.* (*53*) also reported on genetic modifications to the PHA enzyme system from *A. eutrophus*. All three genes from *A. eutrophus* were put into *E. coli* in the sequence: *phaC*, *phaA*, *phaB*. The synthase was overexpressed and purified to greater than 90% homogeneity.

P(3HB) granules have also been reported *in vitro* (*21*). The purified PHA synthase from *A. eutrophus* is combined with chemically synthesized (*R*)-3-hydroxybutyryl-CoA. Granules then form within minutes and nothing further than monomer and synthase are required for granule formation. The granules are spherical with diameters up to 3 μm (*in vivo* granules are approximately 0.5 μm). The molecular weight of the P(3HB) formed is greater than 10×10^6 Da as measured by GPC and this can be controlled by the initial PHA synthase concentration. The kinetics of the reaction show a lag time for the synthase which suggests the involvement of an additional granule assembly step (21). Once formed, the granules tend to coalesce. The authors suggest that, within the time frame of the experiment, the granules are composed of noncrystalline P(3HB). The lag was eliminated with the addition of Hecameg, which indicated that it could not be a priming step (21). The lag was probably due to a granule or micelle assembly step. The granule bound synthase has a higher activity than the soluble form, so this may be evidence for this. The authors suggest that the synthase is not capable of chain transfer to a new chain *in vitro*, as has been suggested for *in vivo* (21, 42).

Recombinant *P. putida*, with the PHA synthase of *Thiocapsa pfennigii*, grown on octanoate produces a 3HB-3-hydroxyhexanoate-3-hydroxyoctanoate copolymer. This is interesting because *T. pfennigii* only synthesizes SCL-PHAs by itself. Thus, a different substrate specificity of the PHA synthase may be revealed when the protein is expressed in different hosts, indicating that bacterial metabolism may impose constraints on the substrates available to the synthase (8).

The exact mechanism of the enzymatic polymerization of long chain *n*-alkanoic acids is unknown, but it is possible that high yields of some MCL-PHAs resulted from higher binding constants of the corresponding monomer units to the synthase, thus leading, in turn, to higher propagation rates relative to chain transfer or chain termination. However, experimental results don't support this, especially in terms of the molecular weights of the polymers which are almost always lower for MCL-PHAs than for SCL-PHAs (41). *P. oleovorans* is able to produce different PHAs with units ranging from 6 to 11 carbons, all of which were copolyesters, none of which had 4 or 5 carbon repeats.

Biodegradation of PHA

One of the strategic properties of PHAs and a source of interest in them is that they are biodegradable plastics. Plastics make up roughly 10 % by weight and about 30 % by volume of all waste material produced in the industrialized and developing world (55). However, there are relatively few countries, Germany, the Netherlands and Belgium, among them, which have a composting infrastructure (25).

There are three types of degradation: chemical, physical or biological (9). Biological degradation is either septic or aseptic. Aseptic biodegradation means that the system is sterile with no microbial activity. Septic biodegradation implies microbial activity, such as is found in soil, compost or lake water. Many PHA degrading organisms are found in the soil. Marine and lake environments are less well studied and there are only a few known PHA degrading organisms in freshwater systems. Both Gram-positive and Gram-negative organisms degrade PHA. There is also wide PHA degrading activity amongst fungi. Depolymerases from *A. faecalis*, *Comamonas testosteroni*, *P. lemoignei*, *P. pickettii* and *P. stutzeri* are also able to degrade homopolymer of P(4HB) (9).

Some general conclusions regarding biodegradation of PHAs are as follows (1). Firstly, the higher the degree of crystallinity, the slower the degradation. Bacterial PHA of 100% (*R*) stereoconfiguration is a preferred substrate over any synthetic PHA structure. Synthetic PHAs degrade more slowly than bacterial PHA and, within the synthetic PHAs, the relative rates are weighted by the degree of crystallinity and the stereoregularity of the polymer. Biodegradability can be adjusted by the use of blends of bacterial PHA and low tacticity synthetic PHA. Most studies of biodegradability involve extracellular PHA depolymerases secreted by bacteria or fungi. Depolymerases can degrade *exo* from the chain end to remover monomer units or oligomers or *endo* by chain cleavage (*1*). Depolymerases can also synergistically combine the two actions. There are no three dimensional structures known of PHA depolymerases, but the primary amino acid sequences are known for a number of them. Intracellular PHA depolymerases are not well studied.

For purpose of this review, blends of PHAs with other polymers have not been considered, although there has been considerable research in this area. In addition, primarily the degradation of SCL-PHAs, especially P(3HB) and P(3HB-*co*-3HV), are considered here.

Intracellular Degradation. Intracellular depolymerases are unable to hydrolyze extracellular PHA and extracellular depolymerases do not hydrolyze intracellular PHA granules (*56-58*). The relationship of the intracellular depolymerases to the other components of the inclusion, in terms of both function and localization, are not well understood (*43*). The substrate specificity of the intracellular depolymerase, as well as the enzyme's properties relative to those of the extracellular depolymerase, are also not known.

Merrick looked at a cell free system of *B. megaterium* granules with soluble depolymerases from *R. rubrum* (*3*). The requirements for depolymerization seem to include a labile factor associated with the granules and the following soluble components: a heat stable protein factor (activator), PHB depolymerase and a hydrolase. 3-hydroxybutyrate is the major product of degradation with about 15 to 20% of 3HB dimer. The hydrolase degrades dimer and trimer (but not P(3HB)) to monomer. The action of the activator precedes the depolymerase and may involve the removal of some protective substance. If granules are treated with mild alkali they no longer need activation (*3*). P(3HB) granules are associated with an inhibitor, presumably a protein which interferes with the hydrolysis of P(3HB) by the depolymerase.

P(3HB) degradation does not involve CoA esters (*15*). *A. beijerinckii* depolymerase, which is associated with the granule membrane, degrades P(3HB) to 3HB which is then oxidized to acetoacetate by an NAD-specific dehydrogenase. Acetoacetate is converted to acetoacetyl-CoA by succinate dehydrogenase (i.e., PHB metabolism is a cyclic process). In both *A. beijerinckii* and *A. eutrophus*, there is a succinate CoA transferase. In *Zoogloea ramigera*, acetoacetate is esterified with CoA by a CoA-synthase using ATP (*15*).

The intracellular depolymerization of P(3HB) in *A. eutrophus* was studied (*59*). P(3HB) granules in *A. eutrophus* released R-(-)-3-hydroxybutyrate when incubated at 37°C. There were two peaks of activity, one at pH 6-7 and the other around pH 8-9. The depolymerase activity was found in the supernatant after treating the granules with protease. Soluble PHB depolymerase activity increased about 20-fold during P(3HB) synthesis. There are possibly two types of depolymerase which are distinct in their effect of ionic strength and pH on P(3HB) hydrolysis. The coupling of PHB depolymerase activity with P(3HB) synthesis may be useful for cells to rapidly adapt to the need for a carbon source through P(3HB) depolymerization (59). The depolymerase is a serine esterase.

The intracellular degradation of P(3HB) granules in *Zoogloea ramigera* I-16-M was reported by Saito *et al.* (60). Release of 3HB was only observed above pH 8.5. When protease treated P(3HB) granules from *A. eutrophus* were used, PHB depolymerase activity was detected in the supernatant fraction of *Z. ramigera*. This is in contrast to when only the *Z. ramigera* depolymerase was used and no activity was detected in the supernatant. The soluble PHB depolymerase showed similar properties to the one found in the granules. The crystallinity and presence of proteins in the granules affect their degradability. The authors suggested that the soluble PHB depolymerase is probably identical to 3HB oligomer hydrolase, which has been purified from *R. rubrum*, *P. lemoignei* and *Z. ramigera* (60). The role of the oligomer hydrolase is not clear since the PHB depolymerase produced 3HB monomer. In addition, the regulation of depolymerization is not known.

Poly(3-hydroxyoctanoate), PHO, inclusions in *P. oleovorans* contain a depolymerase of approximately 32 kDa (43). Association of the depolymerase with the granule is under metabolic regulation relative to the polymerase. The depolymerase has a deduced amino acid sequence similar to lipase.

Extracellular Degradation. The biodegradation of P(3HB) and P(3HB-*co*-3HV) copolymers was studied in soils, under laboratory conditions, *in situ* in waters and in composts by Mergaert and colleagues (61). The influence of time, environment, incubation temperature and %HV were examined. The degrading organisms were isolated and identified. 325 microorganisms (154 bacteria, 77 *Streptomycetes*, 94 moulds) were isolated. The major species obtained included *Acidovorax facilis* and *Variovorax paradoxus* (both Gram-negative bacteria), *Bacillus megaterium* (Gram-positive bacterium) and *Aspergillus fumigatus* (mould). It is questionable as to whether *Alcaligenes faecalis* and *Pseudomonas lemoignei* play an important role in the *in vivo* biodegradation of PHAs (61).

The rate of degradation of PHAs, even under controlled environmental conditions, is difficult to predict. Higher temperatures lead to better degradation, probably due to increased microbial activity. The P(3HB-*co*-3HV) copolymer tended to degrade faster than the homopolymer (61). PHAs were degraded by a wide variety of microorganisms, including Gram-negative bacteria, Gram-positive *Bacilli* and *Streptomycetes*, as well as molds. The relative ratios of degradation of P(3HB) and P(3HB-*co*-3HV) are highly dependent on the environments used and can be quite different from results in *in vitro* tests. In *in vitro* tests, single enzymes are used, while, in natural environments, the composition of degrading microflora and their enzymes varies.

Abe and Doi reported the degradation of synthetic racemic P(3HB) films with isotactic diads between 88% and 54% (with the crystallinity of the films from 33 to

8%) (*62*). Films of syndiotactic P(3HB) (isotactic diads of 30% and 26% crystallinity) were also used. The enzyme was the depolymerase from *A. faecalis*. The degradation was highest for 74% isotactic diads. The syndiotactic film was hardly hydrolyzed. Mixtures of monomer, dimer, trimer and tetramer were obtained, indicating selective hydrolysis of the ester bond connecting the methine carbon. The PHB depolymerase adsorbed on isotactic films, but there was little absorption on the syndiotactic films. This suggests that the binding domain of PHB depolymerase has an affinity toward the isotactic crystalline P(3HB) phase (*62*).

Jesudason *et al.* degraded synthetic P[(*R*,*S*)-3HB] with the extracellular depolymerase from *A. faecalis* T1 (*63*). The high isotactic diad fraction (88%) showed little degradation over 50 hours. The intermediate diad fraction (63%) degraded continuously, but slower than for bacterial P(3HB). The low isotactic fraction (48%) rapidly degraded initially, but then showed no further degradation. The authors reported that the stereoblock structure of the polymer influenced the susceptibility towards degradation (*63*).

Films of synthetic isotactic and syndiotactic P(3HB) were degraded with *P. lemoignei* and *A. fumigatus* M2A depolymerases (*64*). Bacterial P(3HB) was used as a reference. The weight loss of the films was greatest for the bacterial sample (with complete degradation in 48 hours). Both enzymes had similar trends, although the fungal depolymerase was less rapid than the bacterial one. The samples of intermediate tacticity (55 - 60 % isotactic diads) had the greatest weight loss, while the syndiotactic (34 - 45 %) samples were the slowest. Isotactic (68 - 88 %) samples were intermediate. The results were influenced by variations in the crystallinity and stereochemistry of the samples. The *P. lemoignei* depolymerase may be less sensitive to the presence of (*S*) repeat units. The authors proposed that both enzymes require at least a (*R*)-diad in order to catalyze hydrolysis (*64*).

The action of PHB-depolymerase A (from *P. lemoignei*) on compression molded bacterial P(3HB) films was studied by Tomasi *et al.* (*65*). The extent of degradation increased with increasing depolymerase concentration and reached a plateau at ≥ 3 µg/mL. Different thermal treatments were performed on the films to study their effect on degradation at a constant crystallinity. The rate of degradation decreased with an increase in the average size of the crystals (i.e., the perfection of the crystalline phase), although the amorphous phase was preferentially degraded. An increase in the dimensions of the spherulites implies a reduction of the amount of disordered material, which is preferentially degraded (*65*).

Lamellar single crystals of bacterial P(3HB) and synthetic P[(*R*,*S*)-3HB] of various tacticities were degraded with depolymerases from the fungus *Aspergillus fumigatus* and the bacterium *Pseudomonas lemoignei* (*66*). The bacterial P(3HB) single crystals degraded completely with no observed decrease in molecular weight for the partially degraded polymer. The synthetic P(3HB) only partially degraded. The most isotactic material degraded the most and the atactic material did not degrade. These results are different from those for P(3HB) films. However, single crystals normalize the crystallinity of the samples by providing equal exposure to crystalline domains. Degradation was suggested to occur from the crystal edge, rather than from the chain folds of the lamellar surface (i.e., an *exo* mode of action since the molecular weight did not decrease). The syndiotactic samples were intermediate in degradation rate because the crystals actually contained an isotactic component (*66*). There was also some *endo* action by the depolymerase.

Colloidal suspensions of MCL-PHAs (mainly PHO) were prepared for enzymatic degradation (*67*). Six cultures producing extracellular MCL-PHA depolymerases were isolated. All were pseudomonads or related bacteria. All (except for *Xanothomonas maltophilia*) could produce MCL-PHA and, except for *X. maltophilia*, none could hydrolyze P(3HB). Seven *Pseudomonas* strains were screened and all were negative for MCL-PHA depolymerase production. It was concluded that extracellular MCL-PHA depolymerases are found mainly in *Pseudomonas*, but that they are relatively uncommon (i.e., organisms that hydrolyze both SCL- and MCL-PHAs are rare) (*67*).

Comparison of the Action of Extracellular PHA Depolymerases and Lipases. The substrate specificities of extracellular lipases from *Bacillus subtilis*, *Pseudomonas aeruginosa*, *P. alcaligenes*, *P. fluorescens* and *Burkholderia cepacia* (formerly known as *P. cepacia*) and PHA depolymerases from *Comamonas* sp., *P. lemoignei* and *P. fluorescens* GK13 and the esterase from *P. fluorescens* GK13 were studied by Jaeger *et al.* (*68*). All of the lipases and the esterase, but no PHA depolymerases hydrolyzed triolein, indicating a functional difference between lipases and PHA depolymerases. Most lipases could hydrolyze polyesters consisting of a ω-hydroxyalkanoic acid, such as poly(6-hydroxyhexanoic acid), P(4HB), poly(ε-caprolactone) and Bionolle. P(3HB) or other P(3HA)s were not significantly hydrolyzed by the lipases. The PHA depolymerases did not act as lipases in that they could not bind a long chain triacylglycerol or hydrolyze it. The lipase degradation of some PHAs was interesting. It highlights the versatility of lipases with respect to their hydrolytic abilities and may stimulate commercial applications of PHAs (*68*).

The degradation of five PHAs [P(3HB), P(3-hydroxypropionate), P(4HB), P(5HV) and P(6-hydroxyhexanoate)] by 16 lipases and 5 PHA depolymerases was studied by Mukai *et al.* (*69*). Ten eukaryotic lipases (from *Rhizopus* sp., *Penicillum* sp., *Mucor gavanicus*, *Rhizomucor miehei*, *Phycomyces nitens* and *Aspergillus niger*) and six prokaryotic lipases (from *Pseudomonas* sp., *Alcaligenes* sp. and *Chromobacterium vinosum*) were used in addition to PHA depolymerases from *A. faecalis* T1, *Comamonas testosteroni*, *P. lemoignei* depolymerases A and B and *P. pickettii*. None of the lipases degraded films of P(3HB) (*70*), but most degraded P(3-hydroxypropionate) films. Ten lipases degraded P(4HB) films, seven degraded P(5HV) films and four degraded P(6-hydroxyhexanoate) films. The prokaryotic lipases hydrolyzed almost exclusively P(3-hydroxypropionate) films, whereas eukaryotic lipases have a broader substrate specificity and degraded at least two to four types of PHA films. Except for depolymerase B, PHA depolymerases hydrolyzed P(3HB), P(3-hydroxypropionate) and P(4HB). P(5HV) and P(6-hydroxyhexanoate) were not degraded (*69*).

Extracellular Depolymerases. Extracellular depolymerases convert PHAs to water-soluble products. The most studied systems are P(3HB) and P(3HB-*co*-3HV). The first degrading organisms were isolated more than thirty-five years ago and included bacteria of the genuses *Bacillus*, *Pseudomonas* and *Streptomycetes* (*71*). A few years later, Delafield *et al.* (*72*) identified sixteen P(3HB)-degrading bacteria. The number of bacteria able to degrade PHA is not known. However, all known degraders accumulate PHA under growth conditions. PHA-degrading bacteria are divided into eleven groups based on the substrate specificities of their depolymerases (*56*). At least 95 genera of fungi are also known to degrade PHAs.

Delafield *et al.* reported in 1965 that *P. lemoignei* produces extracellular depolymerases which degrade P(3HB) to 3HB monomeric and dimeric esters (*72*). The secretion of the depolymerase occurs mainly at the end of the exponential growth phase. Metabolism of 3HB suppresses enzyme secretion and causes a disappearance of secreted enzyme. The extracellular depolymerase may be 25% or higher of the extracellular protein. Also, 3HB seems to suppress secretion of the depolymerase, which indicates that the depolymerase production may be subject to catabolite repression. It is possible that the depolymerase is reabsorbed by the cells or that it is destroyed at the cell surface (*72*).

All PHA depolymerases share certain characteristics. They are all stable to pH, temperature and ionic strength. They have relatively low molecular masses (< 100 kDa). They are comprised of single peptides. Their optimum pH is in the alkaline range (7.5 to 9.8) (with the only exceptions being the depolymerases of *P. pickettii* and *P. funiculosum* which have optimum pHs of 5.5 and 6.0, respectively), and most appear to be serine hydrolases which are specific for either SCL-PHA or MCL-PHA (*56*). Most bacterial depolymerases are not glycosylated, although eukaryotic and those of *P. lemoignei* are. Glycosylation is not essential for activity, but may assist in the stability of the enzyme (*56*).

All PHA depolymerases are specific for the (*R*) conformer (i.e., P[(*S*)-3HB] does not degrade), but isotactic P[(*R,S*)-3HB] is degraded, although more higher oligomers are obtained. Atactic P(3HB) degrades slowly and, presumably, PHA depolymerases are not able to hydrolyze the ester bonds between monomers of the (*S*) configuration (*56*).

Depolymerase production is usually repressed in the presence of a soluble carbon source (*56*). After exhaustion of nutrients, PHA depolymerase production begins. Very high levels of PHA depolymerase are produced by *P. lemoignei* during growth on succinate after the pH of the medium has become alkaline.

Ten PHA depolymerase structural genes (*phaZ*) have been cloned and analyzed since 1989 (*56*). Most of these have been from Gram-negative bacteria and have been mainly for SCL-PHA (only *P. fluorescens* is specific for MCL-PHA). All of the PHB depolymerases examined have a composite domain structure (see Figure 10). The primary structure of PHA depolymerases are between 393 and 488 amino acids (*73*). There is a 25 to 37 amino acid signal peptide, which is cleaved during passage through the plasma membrane. There is a large catalytic domain near the N-terminus, a substrate-binding domain at the C-terminus and a linker region. Types

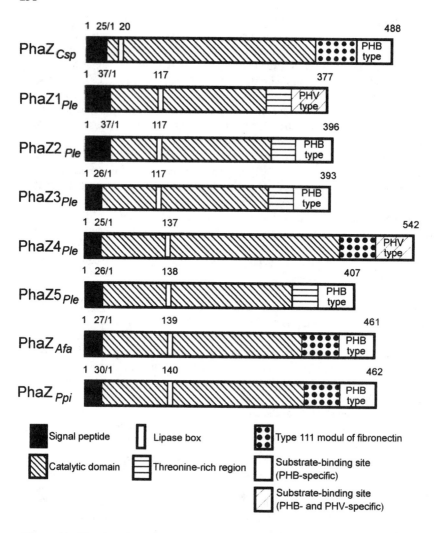

Figure 10. The domain structure of PHA depolymerases. The first and last amino acids of the signal peptides and of the mature protein, and the position of the lipase box serine, are indicated. PhaZ$_{Csp}$ = *Comamonas* sp. depolymerase; PhaZ1$_{Ple}$, PhaZ2$_{Ple}$, PhaZ3$_{Ple}$, PhaZ4$_{Ple}$, and PhaZ5$_{Ple}$ = *Pseudomonas lemoignei* depolymerases; PhaZ$_{Afa}$ = *Alcaligenes faecalis* depolymerase; PhaZ$_{Ppi}$ = *Pseudomonas pickettii* depolymerase. (Reproduced with permission from ref. 74. Copyright 1995 National Research Council of Canada.)

I and II have a Fibronectin type III (Fn3) domain as the linker and type III (i.e., *P. lemoignei*) has a linker rich in threonine (*73*). All active site regions contain strictly conserved serine, aspartate and histidine residues. The serine is part of the "lipase box", a five amino acid sequence (Gly-X1-Ser-X2-Gly) which is found in all serine hydrolases, such as lipases, esterases, and serine proteases (*56, 73*). The oxygen of the serine side chain is the nucleophile which attacks the ester bond and is supported by the imidazole ring of the histidine. The positive charge of the histidine is stabilized by the carbonyl of the aspartate. In most bacterial lipases, X1 is histidine, whereas, in all SCL-PHA depolymerases, it is a leucine. Likely, the hydrophobic side chain of leucine is more compatible with PHAs than the positively charged imidazole ring (*56*).

SCL-PHA depolymerases bind specifically to P(3HB) granules. If the C-terminal domain is removed, this ability is lost and the modified enzymes do not then hydrolyze P(3HB) (*75*). The C-terminal domain is, thus, suggested as a binding domain (*76*). This leads to a proposed two-step reaction for degradation, involving adsorption followed by hydrolysis. Experiments using P(3HB) single crystals suggest that the adsorption is to a crystalline surface (*66*).

The function of the linker is not known. In depolymerases from *P. lemoignei* (i.e. PhaZ1, PhaZ2, PhaZ3 and PhaZ5), it is about 40 amino acids long with a very high proportion of threonine (*56*). In PhaZ4 and all other SCL-PHA depolymerases, this threonine-rich region is replaced by Fibronectin type III (Fn3). It is possible that the linker is just a distance separator between the catalytic and binding domains.

While most bacteria have a single PHA depolymerase, *P. lemoignei* has five SCL-PHA depolymerase structural genes (*56*). These are not clustered, except for *phaZ5* (PHB-depolymerase A) and *phaZ2* (PHB-depolymerase B). Both of these genes are separated by an open reading frame.

The *P. lemoignei* PHA depolymerase system has been well studied (*77*). Five different PHA depolymerase structural genes (*phaZ1* to *phaZ5*) have been identified which encode the enzymes PHB-depolymerase C, PHB-depolymerase B, PHB-depolymerase D, PHV-depolymerase and PHB-depolymerase A, respectively. The five PHA depolymerases from *P. lemoignei* were compared to each other and to the PHA depolymerases from *A. faecalis* and *P. pickettii* (*77*). All of the PHA depolymerases had a domain structure with 25 to 83 % homology to each other. At their N-terminus, they all have typical signal peptides of exoenzymes. As well, all PHA depolymerases have the lipase box, a conserved histidine and aspartate, the catalytic triad and a conserved region similar to the oxyanion hole in lipases (*77*). In PhaZ1, PhaZ2, PhaZ3 and PhaZ5, there is a threonine-rich region (22 to 27 of 40 amino acids) at the C-terminus of the catalytic domain. In PhaZ4 and the depolymerases from *A. faecalis* and *P. pickettii*, there is a 90 amino acid sequence resembling Fn3 of eukaryotic extracellular matrix proteins. The function of the Fn3 sequence in the depolymerase is unknown.

There were two types of sequences at the C-terminus of the depolymerases which represent substrate binding sites (*77*). A PHB-type site was present in the depolymerases of *A. faecalis* and *P. pickettii* and in PhaZ2, PhaZ3 and PhaZ5. A PHV-type site was present in PhaZ1 and PhaZ4. All of the PHA depolymerases hydrolyzed P(3HB). PhaZ1 and PhaZ4 also cleaved P(3HV)and PhaZ4 additionally cleaved P(4HB). All wild type depolymerases were glycosylated, although this was not required for activity.

PHB-depolymerase A and PHB-depolymerase B have molecular weights of 55 and 67 kDa, respectively, and hydrolyze both P(3HB) and P(3HB-co-3HV) (78). PHV-depolymerase has a molecular weight of 54 kDa and hydrolyzes PHV (79). The N-terminal sequences for PHB-depolymerases A, B and C and PHV-depolymerase were determined and compared to the PHB-depolymerase of *Comamonas* sp. *phaZ3* codes for a novel PHB depolymerase of *P. lemoignei*, PHB-depolymerase D. The genes *phaZ1 to phaZ3* have high (68 to 72 %) homology to each other and medium homology to the *A. faecalis* T1 depolymerase (25 to 34 %). All of these depolymerases had the lipase box, a threonine-rich region near the C-terminus, the catalytic triad of Asp, His and Ser (like lipases) and all were glycosylated *in vivo* (78, 80).

Saito *et al.* (81) examined the extracellular depolymerase from *Alcaligenes faecalis*. *A. faecalis* T1 is a Gram-negative bacteria which can hydrolyze water-insoluble P(3HB) and water-soluble 3-HB oligomeric esters. This depolymerase was cloned, sequenced and expressed in *E. coli*, where it was fully active. The recombinant *E. coli* secreted a fraction of the depolymerase into the culture medium, but this was not due to rupturing of the cells (81).

The extracellular PHB-depolymerase from *A. faecalis* T1 was isolated from activated sludge and cultures grown on P(3HB) as the sole carbon source (82). The enzyme is 48 kDa and has an optimum pH of 7.5. The depolymerase degraded P(3HB) from *Z. ramigera* I-16-M, but did not attack native P(3HB) granules. The depolymerase commenced degradation at the free hydroxyl end and released 3HB dimer. It lost its PHB-degrading ability after mild trypsin treatment, but retained oligomer activity. The trypsin-treated depolymerase seemed less hydrophobic, which suggests a hydrophobic binding site that is removed by trypsin (75). Shirakura *et al.* determined that this depolymerase has two disulfide bonds, one of which seems necessary for full enzyme activity (83).

A. faecalis PHB-depolymerase cleaves mainly the second or third ester linkage from the free hydroxyl end of the polymer chain (i.e., acts in an *exo* manner). It also hydrolyzes cyclic oligomeric hydroxybutyryl esters, so it can also act in an *endo* fashion. Typically, monomers to trimers are obtained as products of the degradation (84). Afterwards, oligomers are hydrolyzed to monomers by oligomer hydrolases (85-87).

Five extracellular depolymerases were purified and compared to that from *A. faecalis* T1 (88). All were similar to the one from *A. faecalis* and they could be divided into 3 groups based on their amino acid sequences. The PHB depolymerases were from Gram-negative rod shaped bacteria and were isolated from river water, farm soil, mountain soil and activated sludge. All five depolymerases were between 40 and 50 kDa. All had serine residues in the active site and an essential cysteine bond formed. All were structurally similar to that from *A. faecalis* T1 (88).

The kinetics of adsorption and hydrolysis by extracellular PHB-depolymerase from *A. faecalis* on the surface of five types of PHA films was examined by Kasuya *et al.* (*73*). P(3HB), poly(3-hydroxypropionate) and P(4HB) films were hydrolyzed by the enzyme, but films of poly(*S*-2-hydroxypropionate) and poly(6-hydroxyhexanoate) were not degraded. PHB-depolymerase adsorbed on all of the films and the adsorption kinetics followed the Langmuir isotherm. This is taken as evidence that the binding domain of the depolymerase is not specific for PHA films, whereas the active site is specific for the hydrolysis of PHA films.

The gene for the extracellular P(3HO) depolymerase (*phaZ$_{Pfl}$*) from P. *fluorescens* GK13 was identified by Schrimer and Jendrossek (*89*). The depolymerase has 278 amino acids and is approximately 27 kDa in size. This is a strongly hydrophobic protein. It contains the lipase box and is a glycosylated serine hydrolase. Replacement of serine-172 with alanine inactivated the enzyme (*89, 90*). This depolymerase has low sequence homology (20 to 40%) with extracellular PHA depolymerases from *P. lemoignei* and *A. faecalis* as well as intracellular P(3HO) depolymerases from *P. oleovorans* and *P. aeruginosa*. The differences between the intracellular and extracellular depolymerases are due to different localizations and different physical structures. PhaZ$_{Pfl}$ is a true extracellular enzyme which acts on crystalline PHO granules, whereas intracellular PHO depolymerases presumably act on native, amorphous granules (*89*).

The PHB-depolymerase structural gene of *Comamonas* sp. (*phaZ$_{Csp}$*) was cloned into *E. coli* by Jendrossek *et al.*(*74*). This PHB-depolymerase has 514 amino acids and is 53 kDa. The catalytic domain is approximately 300 amino acids, with a Fn3 sequence. It possesses a PHB-type binding domain. In addition the catalytic triad of serine, histidine and aspartate is present, as is an oxyanion hole structure.

Bacteria which grow on P(3HB) as the sole carbon source were isolated by Jendrossek *et al.* from various soils, lake water and activated sludge (*91*). Although all of the bacteria could grow on a variety of substrates, most could only degrade P(3HB) and P(3HB-*co*-3HV). Five strains could also degrade PHV. PHO was not degraded by any of these strains. One strain, *Comamonas* sp., was selected and the depolymerase purified. This was a nonglycosylated protein of 45 kDa. The depolymerase hydrolyzed P(3HB), P(3HB-*co*-3HV) and (very slowly) PHV. A different mechanism from the depolymerases of *P. lemoignei* or *A. faecalis* T1 was postulated since the *Comamonas* depolymerase produced mainly monomeric 3HB, whereas *P. lemoignei* and *A. faecalis* T1 produce mainly dimer and trimer (*91*).

Comamonas testosteroni was isolated from soil by Kasuya *et al.* (*92*). The strain *C. testosteroni* ATSU grew on P(3HB) as the sole carbon source and secreted an extracellular depolymerase which was also secreted in the presence of P(3HB-co-3HV). The depolymerase was a protein of 49 kDa with optimum activity at pH 8.5. It produced monomers and dimers of 3HB. It had stronger adsorption onto films of P(3HB) than the depolymerases from *A. faecalis* or *P. pickettii*.

C. testosteroni was also isolated from sea water by the same authors (*93*). It grew on P(3HB) as the sole carbon source and secreted a PHA depolymerase. This enzyme was 50 kDa and had optimum activity between pH 9.5 and 10.0. It showed good adsorption on P(3HB) films and degraded P(3HB-*co*-3HV) and P(3HB-*co*-4HB) films faster than P(3HB). An excess of the depolymerase actually completely

142

inhibited the degradation. The measured hydrophobicity of PHA depolymerases was greater for *C. testosteroni* than for *A. faecalis* T1 or *P. lemoignei* depolymerase A. The same authors then isolated ten species of Gram-negative PHA-degrading bacteria from fresh water (*94*). The hydrophobicities of the depolymerases were determined on the basis of partitioning between hydrophobic gels and buffers. The PHA depolymerase from *P. stutzeri* YM1414 is 48 kDa with optimum activity at pH 9.5 and 55 °C. This depolymerase has a rate of enzymatic degradation of P(3HB) films about twice as high as depolymerases from *A. faecalis*, *P. pickettii*, *C. testosteroni* and *P. lemoignei*. The hydrophobicities are higher for depolymerases from aqueous environments than for those from sludge or soil (*94*).

Literature Cited

1. Marchessault, R. H. *Trends Polym. Sci.* **1996**, *4*, 163.
2. Poirier, Y.; Dennis, D. E.; Nawrath, C.; Somerville, C. *Adv. Mater.* **1993**, *5*, 30.
3. Merrick, J. M. *Polym. Preprints* **1988**, *29* (1), 586.
4. Kawaguchi, Y.; Doi, Y. *FEMS Microbiol. Lett.* **1990**, *70*, 151.
5. Müller, H.-M.; Seebach, D. *Angew. Chem. Int. Ed. Engl.* **1993**, *32*, 477.
6. Lauzier, C.; Revol, J.-F.; Marchessault, R. H. *FEMS Microbiol. Rev.* **1992**, *103*, 299.
7. Jackson, D. E.; Srienc, F. *Ann. N. Y. Acad. Sci.* **1994**, *745* (Biochem. Eng. VIII), 134.
8. Poirier, Y.; Nawrath, C.; Somerville, C. *Bio/Technology* **1995**, *13*, 142.
9. Brandl, H.; Bachofen, R.; Mayer, J.; Wintermantel, E. *Can. J. Microbiol.* **1995**, *41* (Suppl. 1), 143.
10. Lemoigne, M. *C. R. Acad. Sci.* **1925**, *180*, 1539.
11. Lemoigne, M. *Bull. Soc. Chim. Biol.* **1926**, *8*, 770.
12. Lemoigne, M. *Bull. Soc. Chim. Biol.* **1927**, *9*, 446.
13. Bréchot, P.; Desveaux, R. *Ann. Inst. Nat. Agr.* **1967**, *5*, 9.
14. Wallen, L. L.; Rohwedder, W. K. *Environ. Sci. Technol.* **1974**, 576.
15. Dawes, E. A. *Bioscience Rep.* **1988**, *8*, 537.
16. Steinbüchel, A.; Valentin, H. E. *FEMS Microbiol. Lett.* **1995**, *128*, 219.
17. John, M. E.; Keller, G. *Proc. Natl. Acad. Sci. USA* **1996**, *93*, 12768.
18. Haywood, G. W.; Anderson, A. J.; Williams, D. R.; Dawes, E. A. *Int. J. Biol. Macromol.* **1991**, *13*, 83.
19. Steinbüchel, A.; Debzi, E. M.; Marchessault, R. H.; Timm, A. *Appl. Microbiol. Biotechnol.* **1993**, *39*, 443.
20. Yamane, T.; Chen, X.-F.; Ueda, S. *Appl. Environ. Microbiol.* **1996**, *62*, 380.
21. Gerngross, T. U.; Martin, D. P. *Proc. Natl. Acad. Sci. USA* **1995**, *92*, 6279.
22. Spyros, A.; Marchessault, R. H. *Macromolecules* **1995**, *28*, 6108.
23. Steinbüchel, A.; Hustede, E.; Liebergesell, M.; Pieper, U.; Timm, A.; Valentin, H. *FEMS Microbiol. Rev.* **1992**, *103*, 217.
24. Inoue, Y.; Yoshie, N. *Prog. Polym. Sci.* **1992**, *17*, 571.

25. Williams, S. F.; Peoples, O. P. *CHEMTECH* **1996**, *26* (9), 38.
26. Hocking, P. J.; Marchessault, R. H. In *Chemistry and Technology of Biodegradable Polymers;* Griffin, G. J. L. Ed.; Blackie Academic & Professional, Glasgow, **1994**; Chapter 4.
27. Harrison, S. T. L.; Chase, H. A.; Amor, S. R.; Bonthrone, K. M.; Sanders, J. K. M. *Int. J. Biol. Macromol.* **1992**, *14*, 50.
28. Ratner, B. D.; Hoffman, A. S.; Schoen, F. J.; Lemons, J. E. *Biomaterials Science: An Introduction to Materials in Medicine*; Academic Press, New York, **1996**.
29. Okamura, K.; Marchessault, R. H. In *Conformation of Biopolymers*; Ramachandran, G. N. Ed.; Academic Press, London, **1967**; Vol. 2, 709.
30. Cornibert, J.; Marchessault, R. H. *J. Mol. Biol.* **1972**, *71*, 735.
31. Yokouchi, M.; Chatani, Y.; Tadokoro, H.; Teranishi, K.; Tani, H. *Polymer* **1973**, *14*, 267.
32. Bruckner, S.; Meille, S. V.; Malpezzi, L.; Cesaro, A.; Navarini, L.; Tombolini, R. *Macromolecules* **1988**, *21*, 967.
33. Marchessault, R. H.; Debzi, E. M.; Revol, J. F.; Steinbüchel, A. *Can. J. Microbiol.* **1995**, *41* (Suppl. 1), 297.
34. Bluhm, T. L.; Hamer, G. K.; Marchessault, R. H.; Fyfe, C.; Veregin, R. P. *Macromolecules* **1986**, *19*, 2871.
35. Gross, R. A.; DeMello, C.; Lenz, R. W.; Brandl, H.; Fuller, R. C. *Macromolecules* **1989**, *22*, 1106.
36. Marchessault, R. H.; Monasterios, C. J.; Morin, F. G.; Sundararajan, P. R. *Int. J. Biol. Macromol.* **1990**, *12*, 158.
37. Marchessault, R. H.; Bluhm, T. L.; Deslandes, Y.; Hamer, G. K.; Orts, W. J.; Sundararajan, P. R.; Taylor, M. G.; Bloembergen, S.; Holden, D. A. *Makromol. Chem. Macromol. Symp.* **1988**, *19*, 235.
38. Ellar, D.; Lundgren, D. G.; Okamura, K.; Marchessault, R. H. *J. Mol. Biol.* **1968**, *35*, 489.
39. Kurja, J.; Zirkzee, H. F.; de Koning, G. M.; Maxwell, I. A. *Macromol. Theory Simul.* **1995**, *4*, 839.
40. de Koning, G. M.; Maxwell, I. A. *J. Environ. Polym. Degrad.* **1993**, *1*, 223.
41. Brandl, H.; Gross, R. A.; Lenz, R. W.; Fuller, R. C. *Appl. Environ. Microbiol.* **1988**, *54*, 1977.
42. Kawaguchi, Y.; Doi, Y. *Macromolecules* **1992**, *25*, 2324.
43. Stuart, E. S.; Foster, L. J. R.; Tehrani, A.; Lenz, R. W.; Fuller, R. C. *Int. J. Biol. Macromol.* **1996**, *19*, 171.
44. Eggink, G.; de Waard, P.; Huijberts, G. N. M. *FEMS Microbiol. Rev.* **1992**, *103*, 159.
45. van der Leij, F. R.; Witholt, B. *Can. J. Microbiol.* **1995**, *41* (Suppl.1), 222.
46. Fukui, T.; Yoshimoto, A.; Matsumoto, M.; Hosokawa, S.; Saito, T.; Nishikawa, H.; Tomita, K. *Arch. Microbiol.* **1976**, *110*, 149.
47. Anderson, A. J.; Williams, D. R.; Dawes, E. A.; Ewing, D. F. *Can. J. Microbiol.* **1995**, *41* (Suppl. 1), 4.

48. Masamune, S.; Walsh, C. T.; Sinskey, A. J.; Peoples, O. P. *Pure Appl. Chem.* **1989**, *61*, 303.

49. Page, W. J.; Manchak, J. *Can. J. Microbiol.* **1995**, *41* (Suppl. 1), 106.

50. Liebergesell, M.; Sonomoto, K.; Madkour, M.; Mayer, F.; Steinbüchel, A. *Eur. J. Biochem.* **1994**, *226*, 71.

51. Wodzinska, J.; Snell, K. D.; Rhomberg, A.; Sinskey, A. J.; Biemann, K.; Stubbe, J. *J. Am. Chem. Soc.* **1996**, *118*, 6319.

52. Gerngross, T. U.; Reilly, P.; Stubbe, J.; Sinskey, A. J.; Peoples, O. P. *J. Bacteriol.* **1993**, *175*, 5289.

53. Gerngross, T. U.; Snell, K. D.; Peoples, O. P.; Sinskey, A. J.; Cushai, E.; Masamune, S.; Stubbe, J. *J. Biochemistry* **1994**, *33*, 9311.

54. Griebel, R.; Merrick, J. M. *J. Bacteriol.* **1971**, *108*, 782.

55. "All that remains: A survey of waste and the environment." *The Economist* May 29, **1993**.

56. Jendrossek, D.; Schrimer, A.; Schlegel, H. G. *Appl. Microbiol. Biotechnol.* **1996**, *46*, 451.

57. Merrick, J. M.; Doudoroff, M. *J. Bacteriol.* **1964**, *88*, 60.

58. Griebel, R.; Smith, Z.; Merrick, J. M. *Biochemistry* **1968**, *7*, 3676.

59. Saito, T.; Takizawa, K.; Saegusa, H. *Can. J. Microbiol.* **1995**, *41* (Suppl. 1), 187.

60. Saito, T.; Saegusa, H.; Miyata, Y.; Fukui, T. *FEMS Microbiol. Rev.* **1992**, *103*, 333.

61. Mergaert, J.; Anderson, C.; Wouters, A.; Swings, J.; Kersters, K. *FEMS Microbiol. Rev.* **1992**, *103*, 317.

62. Abe, H.; Doi, Y. *Macromolecules* **1996**, *29*, 8683.

63. Jesudason, J. J.; Marchessault, R. H.; Saito, T. *J. Environ. Polym. Degrad.* **1993**, *1*, 89.

64. Timmins, M. R.; Lenz, R. W.; Hocking, P. J.; Marchessault, R. H.; Fuller, R. C. *Macromol. Chem. Phys.* **1996**, *197*, 1193.

65. Tomasi, G.; Scandola, M.; Briese, B. H.; Jendrossek, D. *Macromolecules* **1996**, *29*, 507.

66. Hocking, P. J.; Marchessault, R. H.; Timmins, M. R.; Lenz, R. W.; Fuller, R. C. *Macromolecules* **1996**, *29*, 2472.

67. Ramsay, B. A.; Saracovan, I.; Ramsay, J. A.; Marchessault, R. H. *J. Environ. Polym. Degrad.* **1994**, *2*, 1.

68. Jaeger, K.-E.; Steinbüchel, A.; Jendrossek, D. *Appl. Environ. Microbiol.* **1995**, *61*, 3113.

69. Mukai, K.; Doi, Y.; Sema, Y.; Tomita, K. *Biotechnol. Lett.* **1993**, *15*, 601.

70. Brandl, H.; Aeberli, B.; Bachofen, R.; Schwegler, I.; Müller, H.-M.; Bürger, M. H.; Hoffman, T.; Lengweiler, U. D.; Seebach, D. *Can. J. Microbiol.* **1995**, *41* (Suppl. 1), 180.

71. Chowdhury, A. A. *Arch. Microbiol.* **1963**, *47*, 167.

72. Delafield, F.; Doudoroff, M.; Palleroni, N. J.; Lusty, C. J.; Contopoulos, R. *J. Bacteriol.* **1965**, *90*, 1455.

73. Kasuya, K.; Inoue, Y.; Doi, Y. *Int. J. Biol. Macromol.* **1996**, *19*, 35.

74. Jendrossek, D.; Backhaus, M.; Andermann, M. *Can. J. Microbiol.* **1995**, *41* (Suppl. 1), 160.

75. Fukui, T.; Narikawa, T.; Miwa, K.; Shirakura, Y.; Saito, T.; Tomita, K. *Biochim. Biophys. Acta* **1988**, *952*, 164.

76. Kasuya, K.; Inoune, Y.; Yamada, K.; Doi, Y. *Polym. Degrad. Stab.* **1995**, *48*, 167.

77. Jendrossek, D.; Frisse, A.; Behrends, A.; Andermann, M.; Kratzin, H. D.; Stanislawski, T.; Schlegel, H. G. *J. Bacteriol.* **1995**, *177*, 596.

78. Briese, B. H.; Schmidt, B.; Jendrossek, D. *J. Environ. Polym. Degrad.* **1994**, *2*, 75.

79. Müller, B.; Jendrossek, D. *Appl. Microbiol. Biotechnol.* **1993**, *38*, 487.

80. Jendrossek, D.; Müller, B.; Schlegel, H. G. *Eur. J. Biochem.* **1993**, *218*, 701.

81. Saito, T.; Suzuki, K.; Yamamoto, J.; Fukui, T.; Miura, K.; Tomita, K.; Nakanishi, S.; Odani, S.; Suzuki, J.-I.; Ishikawa, K. *J. Bacteriol.* **1989**, *171*, 184.

82. Tanio, T.; Fukui, T.; Shirakura, Y.; Saito, T.; Tomita, K.; Kaiho, T.; Masamune, S. *Eur. J. Biochem.* **1982**, *124*, 71.

83. Shirakura, Y.; Fukui, T.; Saito, T.; Okamoto, Y.; Narikawa, T.; Koide, K.; Tomita, K.; Takemasa, T.; Masamune, S. *Biochim. Biophys. Acta* **1986**, *880*, 46.

84. Nakayama, K.; Saito, T.; Fukui, T.; Shirakura, Y.; Tomita, K. *Biochim. Biophys. Acta* **1985**, *827*, 63.

85. Doi, Y.; Kitamura, S.; Abe, H. *Macromolecules* **1995**, *28*, 4822.

86. Delafield, F. P.; Cooksey, K. E.; Doudoroff, M. *J. Biol. Chem.* **1965**, *240*, 4023.

87. Shirakura, Y.; Fukui, T.; Tanio, T.; Nakayama, K.; Matsuno, R.; Tomita, K. *Biochim. Biophys. Acta* **1983**, *748*, 331.

88. Shiraki, M.; Shimada, T.; Tatsumichi, M.; Saito, T. *J. Environ. Polym. Degrad.* **1995**, *3*, 13.

89. Schrimer, A.; Jendrossek, D. *J. Bacteriol.* **1994**, *176*, 7065.

90. Schrimer, A.; Matz, C.; Jendrossek, D. *Can. J. Microbiol.* **1995**, *41* (Suppl. 1), 170.

91. Jendrossek, D.; Knoke, I.; Habibian, R. B.; Steinbüchel, A.; Schlegel, H. G. *J. Environ. Polym. Degrad.* **1993**, *1*, 53.

92. Kasuya, K.; Doi, Y.; Yao, T. *Polym. Degrad. Stab.* **1994**, *45*, 379.

93. Mukai, K.; Yamada, K.; Doi, Y. *Polym. Degrad. Stab.* **1993**, *41*, 85.

94. Mukai, K.; Yamada, K.; Doi, Y. *Polym. Degrad. Stab.* **1994**, *43*, 319.

Chapter 10

Yeast β-Glucosidases and the Saccharification of Cellulosic Biomass

S. N. Freer, C. D. Skory, and R. J. Bothast

Fermentation Biochemistry Research Unit, National Center for Agricultural Utilization Research, Agricultural Research Service, U.S. Department of Agriculture, 1815 N. University Street, Peoria, IL 61604

Biomass has the potential to serve as a renewable substrate for the production of fuel ethanol and chemicals, but, because of technical problems and cost considerations, this potential is not currently being realized. Our understanding of the organization and function of endo- and exocellulases has increased dramatically in the last few years. However, the hydrolysis of cellobiose and other soluble cellodextrins is still the rate-limiting step in the enzymatic saccharification of cellulose to glucose. This review presents a brief discussion of the structure of cellulose and the enzymology of cellulases. A general discussion of β-glucosidases and a more detailed review of the function, properties, regulation and molecular biology of yeast β-glucosidases is presented.

Non-renewable energy resources (fossil fuels) are, by far, the most exploited form of energy today. The infrastructure of the modern industrial society is built upon oil which, at present production and usage rates, will be exhausted in the next century. In 1994, the United States consumed 2.71×10^{11} gallons of petroleum (748 million gallons per day). About 1.7×10^{11} gallons (64%) were used for transportation fuels, of which 1.15×10^{11} gallons or 42% of the total were used as motor gasoline (State Energy Data Report 1994: Consumption Estimates, Energy Information Administration, Office of Energy Markets and End Use, US Department of Energy, Washington, D.C.). Although great benefits to society are derived from the use of fossil fuels, potential problems, in the form of global warming, environmental pollution and increased atmospheric carbon dioxide, are associated with its use.

One means to help alleviate these problems is to develop viable technologies for the production of energy using renewable fuels and feedstocks. From 1983 through 1993, American farmers produced an average of 7.56×10^9 bushels of corn per year (Agricultural Fact Book 1996, United States Department of Agriculture, Office of Communications, Washington, D.C.). If the total corn crop was converted to fuel ethanol, assuming starch as the substrate and a yield of 2.6 gallons of ethanol per bushel, 1.96×10^{10} gallons of ethanol could be produced. This would meet about 17% of the United States' 1994 motor gasoline needs.

Converting 100% of the United States corn crop into fuel ethanol is obviously unreasonable. However, it exemplifies the need for alternative substrates for the production of fuel ethanol. It is estimated that 1.01×10^9 tons per year of usable biomass exist in the United States (*1*). The cellulosic portion alone represents a potential of about 7.0×10^{10} gallons of fuel alcohol or about 60% of the volume of motor fuel used in 1994. If the hemicellulose fraction of the biomass is also converted to fuel ethanol, then an additional 5.3×10^{10} gallons of ethanol can be produced. Collectively, this represents the volume of all the gasoline fuel used in the United States in 1994. Although the majority of current research is directed toward the production of fuel ethanol, the sugars produced for fermentation could also be used to produce other desirable fuels or chemicals. However, significant technical and economic challenges must be met before the utilization of large volumes of biomass and/or cellulosic wastes becomes feasible.

The utilization of biomass as a fermentative substrate typically entails several general procedures (*i.e.*, feedstock preparation, pretreatment, saccharification, fermentation, product recovery and waste treatment), of which there have been multiple variations developed. Because of the diversity in the physical and chemical properties of the feedstocks, there has not been a universal pretreatment process developed that is optimal for all feedstocks. Currently, the yeast *Saccharomyces cerevisiae* Hansen is the most commonly used organism for the production of fuel ethanol in the United States. However, *S. cerevisiae* has limitations in that it ferments only glucose, sucrose and maltose, while the hemicellulose sugars (xylose and arabinose) remain unused. Therefore, a great deal of research has been directed toward the development of microorganisms capable of fermenting xylose and arabinose. Molecular techniques have recently been utilized for the introduction of ethanol fermentative or sugar utilization genes into various microorganisms, resulting in the development of new, promising organisms that ferment these sugars (*2-4*).

During the last few years, extensive research has characterized the cellulase systems of various bacteria and fungi (*5*). However, less emphasis has been placed on β-1,4-glucosidases, which hydrolyze the products of cellulase (cellobiose and cellodextrins) to glucose. In this paper, we review some of the general characteristics of cellulose, cellulases and β-glucosidases, as well as the specific characteristics of yeast β-glucosidases and their regulatitory mechanisms.

Saccharification of Cellulose

Cellulose. Cellulose is an insoluble, linear polymer of glucose units linked by β-1,4 glucosidic bonds. For microorganisms to utilize cellulose as a carbon source, they must produce extracellular enzymes, free or associated with the cell surfaces, to convert the insoluble polymer to soluble products, which then can be transported into the cell.

Cellulose of varying chain lengths can be categorized as long chained alpha-cellulose, shorter chained beta-cellulose (degree of polymerization (DP) of 15 to 90 glucose units) and short chained gamma-cellulose (DP<15 glucose units). Native cellulose is estimated to have a DP of from 2,000 to 14,000, while commercial wood pulp has a DP of from 650 to 1,500 and newsprint has a DP of about 1000 (6,7). Cellulose chains of more than six glucosyl units are effectively insoluble. Although glucose is the smallest monomeric unit that can be isolated from cellulose, three-dimensional structure analysis of cellulose reveals that cellobiose is the repeating unit.

Cellulose chains have extended, rigid, zigzag shapes that are the consequence of the β-1,4 linkages. Each glucose unit in cellulose has the potential to form three hydrogen bonds with a monomer in an adjacent chain. This results in a highly stable "crystalline" configuration that is anhydrous and recalcitrant to hydrolysis by acids, bases and enzymes. Regions that do not have extensive inter-chain hydrogen bonding are "amorphous" and are more susceptible to hydrolysis. Native cellulose fibrils are bound together by a matrix composed of hemicellulose, lignin and/or pectin.

Cellulase. The cellulase complex consists of two classes of enzymes: 1,4-β-D-glucan 4-glucanohydrolases (endoglucanases; endo-1,4-β-cellulases; EC 3.2.1.4) and 1,4-β-glucan cellobiohydrolases (exoglucanases; cellobiohydrolases; EC 3.2.1.91). Endocellulases act at random on cellulose to generate internal chain breaks, while exocellulases catalyze the release of primarily cellobiose, but also glucose and cellodextrins, from the ends of the β-glucan substrates. Regardless whether the enzyme is an endo- or exocellulase, every hydrolytic event results in the generation of a new reducing sugar group. However, endocellulases, acting randomly, cause a greater reduction in DP per hydrolytic event than do exocellulases. Exocellulases, acting only at a chain end, cause a greater production of free reducing sugars per hydrolytic event than do endocellulases. A third class of enzymes, β-1,4-glucosidases (β-glucosidases; EC 3.2.1.21), acts to liberate D-glucose units from cellobiose and cellodextrins (8-10). Although many authors catagorize β-glucosidases as part of the cellulase complex, to our knowledge, β-glucosidases do not form a physical complex with endo- or exocellulases or utilize, to a significant extent, cellulose as a substrate. However, β-glucosidases do modulate the rate of cellulose hydrolysis by relieving cellobiose inhibition of the cellulases.

Many cellulases contain a catalytic domain and one or more ancilliary domains. A catalytic domain joined to a cellulose-binding domain(s) is frequently observed. The exact role of the cellulose-binding domain is not currently known. However, it has been shown that they are required for efficient hydrolysis of cellulose, but not soluble substrates (5). All cellulolytic enzymes share the same chemical specificity for β-1,4-glycosidic bonds, which they hydrolyze by a general acid-catalyzed mechanism. The reaction takes place such that there is either retention or inversion of the configuration of the anomeric carbon (11, 12). The difference in the specificity between endo- and exocellulases is thought to be the result of the shape of the catalytic domain. Endocellulases have an open active site cleft that can bind to, and act on, the middle of a cellulose chain. On the other hand, the active site of an exocellulase is located in a tunnel formed by long loops in the protein structure (*13-18*). Thus, the cellulose chain appears to snake through the tunnel to the active site and the enzyme is envisioned as acting in a processive manner as it moves along the chain (*ie.*, the enzyme does not release the substrate after each catalytic event).

Certain cellulases complement each other in a synergistic fashion, such that the rate of reaction is greater than the sum of the component activities. Synergy is rarely seen between two endocellulases, but is observed between certain exocellulases and between exo- and endocellulases (*18*). It is reasonably easy to envision how exo- and endocellulases complement each other. However, until recently, all exocellulases were thought to liberate cellobiose from the nonreducing end of cellulose chains. Thus, different exocellulases should compete with each other for a limited number of hydrolysis sites, (the nonreducing chain ends), rather than complement each other to give synergistic hydrolysis. By the early to middle 1990's, Biely and others showed certain exocellulases attack cellooligosaccharides from the reducing ends (*19-21*). Additionally, Gilkes *et al.* (*22*) demonstrated that two exocellulases from *Cellulomonas fimi* attack carboxymethylcellulose at opposite chain ends, thus helping to explain the synergy observed between certain exocellulases.

β-Glucosidases

General Properties. β-Glucosidases occur ubiquitously in animals, insects, plants, fungi and bacteria. They catalyze the hydrolysis of β-1,4 glycosidic linkages in aryl and alkyl β-glucosides, cellobiose and cellodextrins. Certain β-glucosidases are also able to catalyze the hydrolysis of other linkages, such as $(1\rightarrow2)$-β, $(1\rightarrow3)$-β, $(1\rightarrow6)$-β, and $(1\rightarrow4)$-α glucosidic linkages. Because β-glucosidases have been studied from a wide variety of animals, plants and microorganisms, some general characteristics of these enzymes can be noted. Organisms often produce multiple β-glucosidases whose properties may vary dramatically and may be compartmentalized into different cell types or areas of a cell. In humans and mice, acid β-glucosidase is located in the lysosome, while its neutral counterpart is a cytosolic enzyme. Fungi and yeast often produce multiple β-glucosidases that can be either secreted, bound to the cell surface or located in the cytoplasm. β-Glucosidases of human, fungi and dicotyledonous plants are often glycosylated, while those of monocots (maize and sorghum) are not. The substrate specificities of different β-glucosidases vary greatly. Some plant enzymes are specific for the aglycone moiety of their substrate (*23*), while mammalian neutral β-glucosidase hydrolyzes β-glucosides having the glycones β-D-glucose, β-D-galactose, β-D-xylose and α-L-arabinose and the aglycone moieties ρ-nitrophenol, 4-methylumbelliferyl and a variety of alkyl groups (*24*).

Just as the physical properties, location, number of enzymes an organism produces and the substrate specificity of β-glucosidases vary, so do the proposed physiological roles of the enzymes. Plant β-glucosidases are implicated in numerous metabolic events, such as growth, productivity, ripening of fruits, pigment catabolism, and food/feed toxicity-related reactions. One of the more extensively studied reactions in plants is the release of HCN upon the hydrolysis of cyanogenic glycosides by β-glucosidase. Cyanogenesis is reasonably common in the plant kingdom, as it occurs in more than 3,000 plant species belonging to 110 different families. This includes important food sources, such as cassava and lima beans (25). Cyanogenic glycoside hydrolysis only occurs at a significant rate after plant tissues are disrupted by herbivores, fungal or insect attack, or by mechanical means. It is generally believed that the β-glucosidases and their substrates are separated in the intact plant by compartmentalization at either the tissue or subcellular level. This hypothesis was tested in sorghum seedlings and it was demonstrated that the cyanoglycoside, dhurrin, and its catabolic enzymes are located in different tissues (26).

Another extensively studied β-glucosidase is mammalian acid β-glucosidase. In humans, β-D-glycosyl-N-acylsphingosine glucohydrolase (glucocerabiosidase; human acid β-glucosidase) catalyzes the degradation of glucosyl ceramide to glucose and ceramide. Ceramide is further degraded to sphingosine and fatty acid by acid ceramidase. The deficiency of acid β-glucosidase leads to an often lethal inherited disease (Gaucher's disease), in which glucosyl ceramide accumulates in the lysosomes of reticuloendothelial cells. Two unique aspects of human acid β-glucosidase are that it is (1) the only known membrane-bound β-glucosidase and (2) the only β-glucosidase that requires an activator protein (saposin C) to function. The biochemistry, genetics and molecular biology of acid β-glucosidase is being extensively investigated in an effort to develop a cost-effective treatment for Gaucher's disease (27).

A third area of β-glucosidase interest is that of their role and use in the conversion of cellulosic biomass to fermentable sugars. The commercially used cellulase complex of *Trichoderma reesei* Simmons has been extensively studied (for a recent review, see 28). *T. reesei* secretes large amounts of endocellulase and exocellulase, but produces comparatively small amounts of β-glucosidase. When crude *T. reesei* culture filtrates are used to saccharify cellulose, the rate of saccharification decreases rapidly due to glucose (end-product) inhibition of the β-glucosidase (29). Cellobiose (cellodextrins), the product of the endo- and exocellulase reactions, then accumulates and inhibits exocellulase activity and, to a lesser extent, endocellulase activity. Cellobiose is a much more potent inhibitor of the cellulase complex than is glucose (30). When *T. reesei* cellulase preparations are supplemented with fungal culture filtrates rich in β-glucosidase activity, the amount of glucose produced increases significantly and the extent of both avicel and ball-milled cellulose digestion increases by 30% to 50% (29). Similarly, when crude *Clostridium thermocellum* cellulase preparations are supplemented with cloned *Cl. thermocellum* β-glucosidase, the extent of avicel degradation increases by as much as 10-fold (31). Thus, β-glucosidases not only produce glucose from cellobiose/cellodextrins, but also relieve end-product inhibition of endo- and exocellulases by hydrolyzing cellobiose to less inhibitory glucose (29, 32).

The efficiency of cellulase saccharification can also be increased by a process known as simultaneous saccharification-fermentation (SSF) (33, 34). In the case of ethanol production, cellulose is reacted with cellulases in the presence of a glucose-fermenting yeast, such as *S. cerevisiae*. Because glucose is converted to ethanol as it is produced, β-glucosidase, endocellulase and exocellulase activities are not inhibited by glucose accumulation. The SSF process reduces the risk of microbial contamination due to the presence of ethanol and also reduces capital equipment costs. *T. reesei* cellulases are moderately inhibited by ethanol, however. In contrast, ethanol does not appear to inhibit β-glucosidase to any significant extent (*35*). One of the few disadvantages of yeast-based SSF is that the reaction must operate at temperatures around 30°C in order to maintain yeast viability and productivity, instead of at optimal enzyme saccharification temperatures of 45-50°C.

Yeast β-Glucosidases

Substrate Utilization. Yeasts are unable to hydrolyze cellulose, with the exception of some species belonging to the genera *Trichosporon* (*36, 37*). The aerobic assimilation of cellobiose is a common yeast trait that is found in select species of many yeast genera, including *Kluyveromyces, Candida, Dekkera/Brettanomyces, Pichia, Debaromyces, Hanseniospora* and *Saccharomycopsis*. However, the fermentation of cellobiose to ethanol is a relatively rare yeast trait. Of the 308 yeast species known to aerobically assimilate cellobiose as a sole carbon source, only 12 species ferment it (*38*).

In an effort to increase the efficiency of the SSF process, cellobiose-fermenting yeasts were used, instead of *S. cerevisiae*, to effectively increase the amount of β-glucosidase present in the saccharification mixture and also eliminate the end-product cellobiose inhibition of cellulase. Yeasts were screened for their capacity to ferment both glucose and cellobiose (*39, 40*). Two yeasts, *Candida wickerhamii* and *Clavispora lusitaniae* (syn. *Candida lusitaniae*), were observed to ferment cellobiose rapidly. When either of these yeasts is used in SSF of cellulose, 10% to 30% more ethanol is produced than in similar experiments that employ *S. cerevisiae* (*40*), demonstrating the severe end-product inhibition that cellobiose has upon the cellulase complex. Spindler *et al.* (*41*) recently reported that cellobiose-fermenting yeast produce 5% to 16% more ethanol in the SSF of cellulosics than does *S. cerevisiae*. Due to the advantage that cellobiose fermenting yeasts have in the SSF process, there is a renewed interest in yeast β-glucosidases.

The ability of yeasts to ferment cellodextrins was tested since they are also products of the enzymatic saccharification of cellulose. *C. wickerhamii* was the first yeast shown to ferment cellodextrins to ethanol (*42*) and, to date, only one other yeast, *C. molischiana*, is also known to ferment them (*43, 44*). *P. guilliermondii* aerobically assimilates, but cannot ferment, cellodextrins (*44*). All three yeasts produce β-glucosidase activity that is either extracellular and/or cell-bound.

Physiological Role. The physiological role that many of the β-glucosidases play in yeast metabolism is unknown. It is generally assumed that if β-glucosidase activity is detected, in either the cytoplasm or cell culture broths, then the function of this activity must be cellobiose metabolism. However, the β-glucosidases isolated by Fleming and Duerkson (45) and Marchin and Duerkson (46) from *Kluyveromyces* sps. (Table I and II) are only active against the aryl-β-glucosides arbutin, salicin, esculin and p-nitrophenyl-β-D-glucoside (ρNPG). The enzyme that cleaves cellobiose, thus allowing growth of these yeasts on cellobiose, must be another enzyme. Also, the function of the *C. wickerhamii* cytoplasmic β-glucosidase is unknown. This enzyme does not appear to be involved in the metabolism of cellobiose since *C. wickerhamii* is unable to transport cellobiose across its cytoplasmic membrane (47). Kilker *et al.* (48) proposed that yeast cytoplasmic β-glucosidases might be involved in the processing of oligosaccharides during glycoprotein synthesis. For an enzyme to serve a major role in cellobiose metabolism, its activity must account for the *in vivo* specific rate of cellobiose utilization by the yeast. In other words, adequate enzyme must be present to account for the amount of cellobiose metabolized. For example, if a culture utilizes cellobiose at a rate of 1% per day (10g or 29,000 μmole/day/l), then the culture must produce at least 0.02 U/ml of cellobiase activity. Most published reports have neglected to so calculate the physiological significance of the studied β-glucosidase. Furthermore, most yeast β-glucosidases react with the commonly used artificial substrate, ρNPG, from 10- to several hundred-fold more rapidly than with cellobiose. Therefore, when ρNPG is used as the substrate, the above theoretical culture would have to contain at least 0.2 U/ml of activity to minimally account for the level of cellobiose utilization. Additionally, when small amounts of enzyme activity are detected in culture broths, precautions should be taken to ensure that this is not the result of cell lysis. Many papers on yeast β-glucosidases, unfortunately and similiar to the above criticism, did not conduct critical cellular location experiments.

Location and Multiplicity of Enzymes. The β-glucosidases from fifteen yeast species have been purified and characterized. The current taxonomic classification, as well as synonyms, of these yeast species is listed in Table I.

The cellular location and the number of β-glucosidases that yeasts produce varies with each species. Until recently, it was thought that essentially all yeast β-glucosidases were soluble, cytoplasmic enzymes (49). However, as the enzymes of more yeasts were investigated, it became apparent that some β-glucosidases are extracellular and/or cell-bound. In general, the cell-bound enzymes appear to be tightly associated with the cell walls. They are not removed from the cell surface by detergent or osmotic treatments, but are released by protoplasting, even under mild conditions before the protoplasts become osmotically sensitive. Saha and Bothast (50) screened 48 yeast species belonging to the genera *Candida, Pichia, Debaromyces* and *Kluyveromyces* and found that the supernatants of all species had at least some extracellular β-glucosidase activity, although most contained only mU/ml of activity. Of the sixteen yeasts whose β-glucosidases have been purified and characterized (Table II), four produce at least one extracellular β-glucosidase (*C. peltata, Z. bailii, C. molischiana* and *C. wickerhamii*). *H. vineae (51), C. entomophila (52)* and *D. bruxellensis (53)* when grown in medium containing cellobiose produce β-glucosidase activities that are associated with whole

Table I. Yeasts from which β-glucosidases have been isolated and/or cloned.

Organism	Synonyms
Candida entomophila D.B. Scott, van der Walt & van der Klift	
Candida molischiana (Zikes) S.A. Meyer & Yarrow	*Torulopsis molischiana*
Candida peltata (Yarrow) S.A. Meyer, Ahearn & Yarrow Comb. Nov.	*Torulopsis peltata*
Candida wickerhamii (Capriotti) S.A. Meyer & Yarrow	*Torulopsis wickerhamii*
Debaryomyces etchellsii (Kreger-van Rij) Maeda, Y. Yamada, Banno & van der Walt	*Pichia etchellsii*
Debaryomyces polymorphus (Klöcker) Price & Phaff	*Pichia polymorphus*
Dekkera bruxellensis van der Walt	*Dekkera intermedia*
Hanseniaspora vineae van der Walt & Tscheuschner	*Vanderwaltia vineae*; *Koeckeraspora vineae*
Kluyveromyces marxianus (Hansen) van der Walt	*Saccharomyces fragilis*; *Kluyveromyces fragilil*
Kluyveromyces dobzhanskii (Shehata, Mark & Phaff) van der Walt	*Saccharomyces dobzhanskii*
Kluyveromyces lactis (Dombrowski) van der Walt	*Saccharomyces lactis*
Pichia farinosa (Lindner) Hansen	*Candida cacaoi*
Pichia guilliermondii (Wickerham)	*Candida guilliermondii*
Pichia anomala (Hansen) Kurtzman	*?Candida pelliculosa* var. *acetetherius*; *Hansenula anomala*
Rhodotorula minuta (Saito) F.C. Harrison	
Saccharomycopsis fibuligera (Lindner) Klöcker	
Zygosaccharomyces bailii (Lindner) Guilliermond	*Saccharomyces elegans*

Table II. Physical and kinetic properties on yeast β-glucosidases.

Organism	Site[a]	Molecular Mass (kDa)		pI	pH Optima (°C)	CHO (%)	Temp (°C)[b]	Km (mM)		Ki G$_1$ (mM)	Ref.
		Native	Subunit					pNPG	G$_2$[c]		
R. minuta	cyto	300			6.4-6.6			0.08		8.5	67
K. marxianus	cyto	320	90		5.8		45	0.27	64	1.6	95
K. marxianus	cyto	330			5.7-5.8			0.11	NA	3.44	45
K. dobzhanskii	cyto	330			6.1-6.2			0.069	NA	8.47	
K. mar. x K. dob.	cyto	330			5.9-6.1			0.086-1.25	NA		
K. mar. x K. dob.	cyto	315			6.6		50	0.09	NA	8.8	96
K. lactis Y-123	cyto				6.2	None		0.201	NA	3.15	46
Y-14	cyto				6.8			0.157	NA	6.71	
Y-1057A	cyto				6.8			0.105	NA	23.0	
P. anomala	cyto	360	90	4.9	6.5		50	0.5	37	6.5	97
P. farinosa	cyto	220	110	3.6	5.5		50	0.44	90	8.0	55
D. polymorpha	cyto				6.0-6.5		40	0.72	220		98
	cyto				6.0-6.5		40	0.21	400		

H. vineae	cyto	295	79	4.5	6.0-6.5		50	0.1	8.0	6.5	51
P. guilliermondii	cyto?	48			6.8		45	0.125			99
Z. bailii	cyto	440	90	4.5	5.5	9.0	55	0.28	83	0.105	56
	extra	360	110	4.65	5.0	8.4	55	0.18	25	0.137	
C. entomophila	cyto	400	100	3.7	5.0-6.0		45	0.177	55.6	3.1	52
C. molischiana CBS-136	extra	120	64	3.7	4.0-4.5	4.0	55	0.1	130	7.0	54
35	extra			3.6	4.0	60	50	0.24	500	9.5	63
35M5N	extra	380				70	50				72
C. wickerhamii	cyto	130	48		6.0-6.3			0.14	3.0	9.0	66
	cyto		97		7.0-7.4		40	0.185	NA	7.5	100
	cyto	180	94	4.3	6.0-6.5		28	0.28		20	94
	peri	198	94	3.9	4.5	30.5	40	3.13	211	None	58, 60
	extra	130	98, 38		4.5		40	4.0	225	230	66
	extra	151	83	3.2	4.25	12		2.8	300	190	59

a. Cellular location of the purified β-glucosidase; "cyto" is cytoplasmic; "extra" is extracellular; "peri" is periplasm or cell-bound.
b. Highest temperature at which the enzyme is stable.
c. NA represent no activity, blanks represent no data available.

cells (cell-bound) and are in cell macerates.   Culture broths contain no detectable activity.   The presumed cytoplasmic β-glucosidases from these yeasts have been purified, but it is unknown if the enzymes responsible for the whole cell and cell macerate activities are identical.

C. molischiana (54) produces both cell-bound and extracellular β-glucosidase activities, which the authors assume are the same enzymes. No cytoplasmic activity was detectable in C. molischiana.

P. farinosa (55) and Z. bailii (56) produce extracellular, cell-bound and cytoplasmic β-glucosidase activities. The extracellular and cytoplasmic enzymes have been purified and shown to have different physical and kinetic properties (Table II). These authors state that both yeasts produce only two different β-glucosidases; ie., they presume the cell-bound and extracellular enzymes to be identical, although physical and kinetic comparisons were not conducted.

Table III. N-Terminal Amino Acid Sequences of the Three C. wickerhamii β-Glucosidases

Location	N-Terminal Sequence	Ref.
Cytoplasmic	MVGFDIEVKLSILTLDEKAKLIAGTDFXHTH	57
Cell-bound	ADDASKPGIGKFAPGQLGF	61
Extracellular	SVDAHKPGKHKWAWDQIG	59

C. wickerhamii also produces extracellular, cell-bound and cytoplasmic β-glucosidase activities (Table II). Enzymes from all three locations have been purified and characterized. The cytoplasmic enzyme has distinct physical and kinetic properties (57) from the cell-bound and extracellular enzymes, whose physical and kinetic properties are very similar (58-60). However, the amino acid composition, level of glycosylation (Table II) and N-terminal amino acid sequences (Table III) of the three enzymes are different (57, 59, 61), indicating that the cell-bound and extracellular enzymes are not the same, but rather distinct enzymes encoded by separate genes.

Catalytic Activity.   Yeast β-glucosidases display a great range in their substrate specificities.   In general, all yeast β-glucosidases hydrolyze cellobiose, aryl-β-D glucosides (ie., ρNPG, salicin, arbutin or esculin) and alkyl-β-D-glucosides (ie., methyl-β-D-glucopyranoside). The exceptions are enzymes isolated from the *Kluyveromyces* sps. which are inactive on cellobiose (Table II). Yeast β-glucosidases hydrolyze aryl-β-glucosides more easily than cellobiose.   The K_ms for ρNPG range from 0.069 mM to 4.0 mM, while the cellobiose K_ms range from 3.0 mM to 220 mM for the cytoplasmic enzymes and range from 25 mM to 500 mM for the extracellular and cell-bound enzymes. The V_{max}s of the enzymes are usually significantly greater with pNPG than cellobiose, resulting in specific activities that are often 10- to several hundred-fold greater on pNPG.

The extracellular β-glucosidase from *C. peltata* exhibits a strong preference for substrates containing (1→4)-β linkages. The enzyme shows little activity on substrates containing other linkages (*62*). The extracellular *C. wickerhamii* β-glucosidase appears to be reasonably specific for β-linkages and cleaves substrates containing either (1→4)-β, (1→2)-β, (1→3)-β or (1→6)-β linkages (*59*). The *C. wickerhamii* cytoplasmic β-glucosidase, on the other hand, hydrolyzes (1→4)-β, (1→1)-β, (1→4)-β-galactosyl, (1→4)-β-xylosyl and (1→4)-α-linkages (*57*). *H. vineae* (*51*) and *P. farinosa* (*55*) cytoplasmic β-glucosidases have very broad substrate specificities. The *H. vineae* β-glucosidase hydrolyzes substrates containing (1→4)-β, (1→2)-β, (1→3)-β, (1→6)-β, (1→4)-β-galactosyl and (1→4)-α linkages, while the *P. farinosa* enzyme hydrolyzes all the above linkages plus (1→6)-α-rhamnosyl and (1→6)-α-arabinosyl linkages. In addition to hydrolyzing most of the above linkages (*63*), the *C. molischiana* β-glucosidase also hydrolyzes the β-glucoside linkages of monoterpenol glycosides in Muscat grape marc (*64*) and the anthocyanin-β-glucosides in red wine (*65*). It should be noted that the relative activities on substrates having linkages other than (1→4)-β are often less than 5% of that on pNPG.

Cellodextrins of DP 3 to 6 also serve as substrates for the extracellular and cell-bound β-glucosidases of *C. wickerhamii* (*44, 59, 66*) and the extracellular enzymes of *C. molischiana* (*54*) and *C. peltata* (*62*). However, *C. peltata* is unable to ferment cellobiose and it is unknown if cellodextrins can serve as a sole carbon source for growth.

Cellobiose-grown *C. wickerhamii* produces from 3 to 6 U/ml of β-glucosidase activity, of which about 85% is cell-bound and 15% is extracellular (*58*). Both the purified extracellular (*59*) and cell-bound (*60*) enzymes produce glucose from cellobiose and cellodextrins by catalyzing the removal of the terminal glucose moiety from the nonreducing end. Computer analysis can be utilized to estimate the integrated rate equation of cellobiose and cellodextrin utilization by *C. wickerhamii*. When experimentally determined kinetic parameters for the *C. wickerhamii* cell-bound enzyme are used in the computer model, the predicted rates of glucose formation agree well with experimental data. Furthermore, when *S. cerevisiae* is cultured on cellobiose or cellodextrins, and the purified *C. wickerhamii* cell-bound β-glucosidase is added to the *S. cerevisiae* culture at levels that mimick the production of β-glucosidase by a *C. wickerhamii* culture, the two cultures produce ethanol at equivalent rates (*60*). This strongly implies that the role of the cell-bound and extracellular *C. wickerhamii* β-glucosidases is to hydrolyze cellobiose/cellodextrins to glucose and the rate-limiting step in the metabolism of these substrates is the amount of β-glucosidase produced.

Many yeast β-glucosidases are inhibited competitively by β-thioglucosides (*67*) and D-glucono-δ-lactone. β-Thioglucosides are structural analogues of the substrates and inhibit the ground state of the enzymes (*68, 69*). These compounds form complexes with the yeast β-glucosidases, but are not cleaved. D-Glucono-δ-lactone is a transition-state analogue that inhibits essentially all β-glucosidases (70). Glucose also acts as a competitive inhibitor of most yeast β-glucosidases and the K_i is less than 10 mM or 0.18% (w/v). Exceptions are the cell-bound and extracellular β-glucosidases of *C. wickerhamii* and the extracellular *C. peltata* β-glucosidase (Table II). These enzymes might be extremely useful, in conjunction with cellulases, for the generation of glucose streams from cellulosic substrates because of the lack of glucose inhibition.

Physical Properties. Most yeast β-glucosidases are acidic, multisubunit, high molecular weight proteins (Table II). The isoelectric points of these enzymes range from 3.2 to 4.9. The native enzymes isolated from *R. minuta*, the *Kluyveromyces* sps., *C. entomophila*, *P. anomala*, *Z. bailii* and *D. bruxellensis* have molecular masses that range from 300 to 440 kDa. These enzymes are composed of four identical subunits of molecular mass 79 to 110 kDa. The β-glucosidase of *C. molischiana* CBS-136 and *C. wickerhamii* are smaller (120 and 198 kDa., respectively) and are composed of two identical subunits. *P. guilliermondii* and *C. peltata* β-glucosidases appear to be monomers of 48 and 43 kDa., respectively.

Yeast cytoplasmic β-glucosidases typically contain little or no carbohydrate, while the extracellular and cell-bound enzymes are glycoproteins that contain mannose as the predominant, if not sole, neutral sugar. The extracellular *Z. bailii* β-glucosidase, the exception, is reported to contain mainly galactose (56). The *C. wickerhamii* extracellular and cell-bound enzymes contain 12% and 30.5% (w/w) mannose, respectively (58, 59), while the cytoplasmic β-glucosidase contains no detectable carbohydrate (57,66). Many β-glucosidases from filamentous fungi are glycoproteins that contain up to 90% (w/w) carbohydrate (9).

The extracellular β-glucosidase from the type strain of *C. molischiana* (CBS-136) contains 4.0% (w/w) mannose (54). Strain 35 of *C. molischiana* was isolated on agar plates containing cellobiose and 0.05% (w/v) 2-deoxy-D-glucose after exposure to a nonsterile atmosphere (63). The isolate was classified as *C. molischiana*, although it atypically forms a true mycelium with septa, shows slight variations in maltose fermentation, L-arabinose, L-rhamnose and succinic acid assimilation, assimilats DL-lactic acid and produces no β-glucosidase under anaerobic conditions. The extracellular β-glucosidase produced by strain 35 contains 60% (w/w) mannose. After Endo H treatment to deglycosylate the enzyme, its molecular mass was estimated as 100 kDa. by gel filtration and 94 kDa. by denaturing SDS-polyacrylamide gel electrophoresis, indicating that the native enzyme is a monomer (63). We then calculate the molecular mass to be 235 kDa. (94/0.4 = 235). *C. molischiana* strain 35M5N was derived as a derepressed N-methyl-N'-nitro-N-nitrosoguanidine mutant of strain 35 that grows in medium containing cellobiose and 0.12% (w/v) 2-deoxyglucose (71). The extracellular β-glucosidase from this mutant has a native molecular mass of 380 kDa., of which 70% (w/w) is mannose. Upon deglycosylation of the enzyme with Endo H, two bands of molecular mass 89 and 90 kDa. were detected. It is unclear if the two proteins are isoforms of the same enzyme or if they are isoenzymes (two similar enzymes encoded by separate genes) (72). Also, it is unclear why the level of glycosylation varied so dramatically between these three isolates and why the β-glucosidase from the type strain (CBS-136) is a dimer and the enzymes isolated from strains 35 and 35M5N are monomers.

Yeast β-glucosidases are not considered heat-stable, as most denature quite rapidly at temperatures above 45° to 50°C (Table II). Highest temperature stability is exhibited by enzymes from *Z. bailii*, which are stable at 55°C. The most heat-labile is the cytoplasmic *C. wickerhamii* enzyme, which inactivates at temperatures above 28°C. In contrast, several β-glucosidases from filamentous fungi are stable at temperatures of 70°C and above (*73*).

Regulation. The ability of yeast to utilize various carbon sources is highly regulated (*74-76*). For example, sucrase (*77-79*) in *S. cerevisiae* is produced constitutively. However, glucose, fructose and mannose repress its expression (catabolite or glucose repression), resulting in diauxic utilization of glucose and sucrose mixtures. The *S. cerevisiae* maltase is also catabolite repressed, but in the absence of glucose, a maltase induction system is present (*80*).

The expression and regulation of β-glucosidases have been studied in several different yeast species (Table IV). Inducible β-glucosidases are found in *R. minuta*, the *Kluyveromyces* sps. and *P. guilliermondii*. *R. minuta* produces β-glucosidase constitutively, but methyl-β-D-glucopyranoside (βMG), thiomethyl-β-D-glucopyranoside (TMG) and thioethyl-β-D-glucopyranoside (TEG), all at an optimal concentration of 10^{-2} M, increase the rate of synthesis 55- to 70-fold (*81*). Natural substrates such as cellobiose, amygdalin, salicin, arbutin and esculin are poor inducers.

The hybrid yeast *K. marxianus* x *K. dobzhanskii* also produces a constitutive β-glucosidase when grown on succinate synthetic medium. Addition of βMG, TMG or TEG results in a 2- to 3-fold increase in the rate of synthesis. The addition of either arbutin, salicin, esculin or amygdalin (but not cellobiose) causes a 3- to 6-fold induction. Low concentrations of glucose (4 x 10^{-3} M) induce high rates of enzyme synthesis, while higher glucose concentrations lead to repression of constitutive β-glucosidase synthesis. High levels of fructose and mannose also repress β-glucosidase synthesis (*82, 83*). In *K. lactis* strain Y-123, the rate of β-glucosidase synthesis increases 4-fold upon addition of 2 x 10^{-2} M βMG, but no effect is seen with strain Y-14. TMG and TEG have no inductive effect upon Y-123. However, both strains are induced, by as much as 60-fold, by low concentrations (10^{-3} to 10^{-6} M) of glucose. Higher levels of glucose repress β-glucosidase expression. βMG induction is specific for β-glucosidase synthesis, while low concentrations of glucose also induce sucrase, maltase and lactase. This type of expression/regulation, in which the constitutive rate of enzyme synthesis is inducible by βMG or low concentrations of glucose and repressible by higher concentrations of glucose, is termed "semiconstitutive" or "paraconstitutive".

The majority of yeasts are considered to synthesize β-glucosidase constitutively (Table IV), as they produce approximately the same amount of enzyme activity when grown on substrates such as cellobiose, ethanol, xylose or glycerol. Unlike the enzymes whose regulation is paraconstitutive, the rate of β-glucosidase synthesis in these yeasts

Table IV. Expression and regulation of yeast β-glucosidases.

Organism	Expression	Regulation	Ref.
R. minuta	Inducible		*81*
K. marxianus x *K. dobzhanskii*	Inducible	paraconstitutive	*82, 83*
K. lactis Y-123 Y-14	Inducible Constitutive	paraconstitutive	*101*
P. guilliermondii	Inducible	Glucose repressed	*99*
P. furinosa	Constitutive	Glucose repressed	*55*
H. vineae	Constitutive	Glucose repressed	*51*
D. bruxellensis	Constitutive	Glucose repressed	*53*
Z. bailii	Constitutive	Glucose Repressed	*56*
C. peltata	Constitutive	No repression	*62*
C. molischiana CBS-136	Constitutive	$+O_2$, Glucose Repressed; $-O_2$, Glucose Repressed	*54*
35	Constitutive	$+O_2$, Glucose Repressed; $-O_2$, No Activity	*63*
35M5N	Constitutive	$+O_2$, Glucose Repressed	*71*
NRRL Y-2237	Constitutive	$+O_2$, Glucose Repressed; $-O_2$, Glucose Repressed	*102, 103*
C. wickerhamii cytoplasmic	Constitutive	$+O_2$, Glucose Repressed; $-O_2$, Glucose Repressed	*66*
cytoplasmic	Constitutive	$+O_2$, Glucose Repressed	*57*
cell-bound	Constitutive	$+O_2$, Glucose Repressed; $-O_2$, Derepressed	*60*
extracellular	Constitutive	$+O_2$, Glucose Repressed; $-O_2$, Glucose Repressed	*66*

is not induced by βMG or small concentrations of glucose. The synthesis of most of these enzymes are, however, repressed by glucose. Usually, 1% to 5% glucose reduces the amount of β-glucosidase produced by at least 90% and this repression is not reversed by cellobiose or other β-glucosides. An exception to this is the extracellular *C. peltata* β-glucosidase, whose synthesis appears to be resistant to glucose repression (see specific references in Table IV).

Table V. Effect of aeration and carbon source upon growth and β-glucosidase production by *C. wickerhamii*[a].

Carbon Source[b]	Aerobic		Anaerobic	
	A_{600}	β-Gluc. Activity	A_{600}	β-Gluc. Activity
Glucose	17.7	0.0	10.1	5.7
Fructose	18.2	0.0	9.6	6.3
Mannose	20.9	0.0	9.8	5.4
Cellobiose	24.7	3.74	9.9	6.9

[a]100 g of carbohydrate per liter
[b]Data recorded after 2 days of growth

The regulation of the cell-bound and extracellular *C. wickerhamii* β-glucosidases is unique. When grown aerobically, β-glucosidase activity is repressed in the presence of glucose. This results in diauxic utilization of mixtures of glucose and non-repressive carbon sources. When grown fermentatively, glucose repression does not occur. Approximately the same amount of β-glucosidase activity is produced in medium containing either glucose, fructose or mannose, as in medium containing cellobiose (Table V). Mixtures of glucose and cellobiose are co-utilized, even in medium initially containing 90 g of glucose and 10 g of cellobiose per liter. Furthermore, if cultures growing aerobically in medium containing 100 g of glucose per liter (repressed conditions) are shifted to anaerobic conditions, β-glucosidase production commences (*84*). This is the only glucose repressed enzyme we are aware of whose synthesis is derepressed by anaerobiosis. Interestingly, the *C. wickerhamii* cytoplasmic β-glucosidase appears to be glucose repressed under both aerobic and anaerobic growth conditions (*57, 63*).

Expression in *S. cerevisiae*. Because β-glucosidase modulates the rate of saccharification of cellulose and the advantages that cellobiose-fermenting yeasts give to the SSF process, several yeast β-glucosidase genes have been cloned and their expression studied in *S. cerevisiae*, in the hope that such a recombinant organism will impove the efficiency of fuel alcohol production from cellulosic substrates. β-Glucosidase genes have been cloned from six different yeast species, *P. anomala* (*85*), *C. molischiana* (*65, 86*), *K. lactis* (*87*), *P. anomala* (*88, 89*), *S. fibuligera* (*90*) and *C. wickerhamii* (*61*), and their expression in *S. cerevisiae* examined in the latter four. Amino acid comparison of these enzymes with other β-glucosidase sequences suggest those from *P. anomala, S. fibuligera* and *C. molischiana* are members of the family 3 glycosyl hydrolases, while that from *C. wickerhamii* is a member of the family 1 glycosyl hydrolases (*91, 92*).

The *K. lactis* β-glucosidase gene is expressed in the cytoplasm of *S. cerevisiae* at levels of up to 400-times of those produced in the parent yeast. Glucose represses the synthesis of the cloned enzyme, but the *S. cerevisiae* clone is less sensitive than the *K. lactis* parent to glucose repression. The clone is able to assimilate, but not ferment, cellobiose (Kluyver effect), possibly because *S. cerevisiae* cannot transport cellobiose when grown anaerobically (*93*). A *S. cerevisiae* clone, containing a β-glucosidase gene from *P. anomala* (*88, 89*), produces β-glucosidase activity that is located in the periplasmic space. This cloned enzyme is not sensitive to glucose repression and the recombinant *S. cerevisiae* grews on cellobiose. Its ability to ferment cellobiose was not tested. Two different β-glucosidase genes (*bgl1* and *bgl2*) were isolated from *S. fibuligera* and expressed in *S. cerevisiae* (*90*). The secreted cloned enzymes have different substrate specificities. The recombinant Bgl1 hydrolyzes cellobiose, but Bgl2 does not. The *S. cerevisiae* transformant carrying *bgl2* does not ferment cellobiose, while the *bgl1* transformant produces a maximum of 1% ethanol from cellobiose. The gene that encodes the *C. wickerhamii* cell-bound β-glucosidase (*bglB*) was cloned (*61*) and its expression/secretion was studied in *S. cerevisiae* and *Pichia pastoris* (*94*). Various constructs were made using either the *S. cerevisiae adh1* or *gal1* promoters and/or the sucrase secretion signal. Several of the constructs expressed the *bglB* gene and secreted the glycosylated BglB protein. However, only a small portion of the recombinant protein was produced in an active state. Thus, to date, the development of a *S. cerevisiae* clone that efficiently ferments cellobiose and/or produces large amounts of yeast β-glucosidase has eluded researchers.

Literature Cited

1. Detroy, R.W.; St. Julian, G.; Freer, S.N. *Symposium: Chemistry in Energy Production*; OakRidge National Laboratory: Oak Ridge, TN, 1980; pp. 85-98.
2. Ho, N.W.Y.; Tsao, G.T. The Patent Cooperation Treaty (PCT) Patent No. W095/13362, **1995**.
3. Ingram, L.O., Beall, D.S.; Burchhardt, G.F.H.; Guimaraes, W.V.; Ohta, H.; Wood, B.E,; Shanmugam, K.T. U.S. patent 5,424,202, June 1995.
4. Zhang, M.; Eddy, C.; Deanda, K.; Finkelstein, M.; Picataggio, S. *Science* **1995**, *267*, 240-243.

5. Tomme, P.; Warren, R.A.J.; Gilkes, N.R. *Adv. Micro. Physiol.* **1995**, *39*, 1-81.
6. Bisaria, V.S.; Ghose, T.K. *Enzyme Microb. Technol.* **1981**, *3*, 90-104.
7. Katzen, R.; Madson, P.W.; Monceaux, D.A. In *The alcohol textbook: incorporating the alcohol alphebet*; Lyons, T.P.; Kelsall, D.R.; Murtagh, J.E. Ed.; Nottingham University Press: Nottingham, NG11 OAX, United Kingdom, 1995, pp. 37-46.
8. Lee, Y-H., Fan, L.T. *Adv. Biochem. Eng.* **1980**, *17*, 102-129.
9. Coughlan, M.P. In *Biotechnology and Genetic Engineering Reviews*; Russell, G.E., Ed.; Intercept, Ponteland, Newcastle upon Tyne, 1985, Vol. 3; pp. 39-109.
10. Coughlan, M.P.; Ljungdahl, L.G. *FEMS Symp.* **1988**, *43*, 11-30.
11. Sinnott, M.L. *Chem. Rev.* **1990**, *90*, 1171-1202.
12. McCarter, J.D.; Withers, S.G. *Curr. Opin. Struct. Biol.* **1994**, *4*, 885-892.
13. Rouvinen, J.T.; Bergfors, T.; Teeri, T.T.; Knowles, J.K.C.; Jones, T.A. *Science* **1990**, *249*, 380-386.
14. Spezio, M.; Wilson, D.B.; Karplus, P.A. *Biochemistry* **1993**, *32*, 9906-9916.
15. Divne, C.; Stahlberg, J.; Reinkainen, T.; Ruohonen, L.; Pettersson, G.; Knowles, J.K.C.; Teeri, T.T.; Jones, T.A. *Science* **1994**, *265*, 524-528.
16. Meinke, A.; Damude, H.G.; Tomme, P.; Kwan, E.; Kilburn, D.G.; Miller, Jr., R.C.; Warren, R.A.J.; Gilkes, N.R. *J. Biol. Chem.* **1995**, *270*, 4383-4386.
17. Linder, M.; Teeri, T.T. *J. Biotechnol.* **1997**, *57*, 15-28.
18. Teeri, T.T. *Trends Biotechnol.* **1997**, *15*, 160-167.
19. Vrsanská, M.; Biely, P. *Carb. Res.* **1992**, *227*, 19-27.
20. Biely, P.; Vrsanská, M.; Claeyssens, M. In *TRICEL93 Symposium on Trichoderma reesei Cellulases and Other Hydrolayses*; Suominen, P.; Reinkainen, T., Eds.; Foundation for Biotechnical and Industrial Fermentation Research: Espoo, Finland, 1993; pp. 99-108.
21. Barr, B.; Hseih, Y-L; Ganem, B.; Wilson, D. *Biochem.* **1996**, *35*, 586-592.
22. Gilkes, N.R.; Kwan, E.; Kilburn, D.G.; Miller, R.C.; Warren, R.A.J. 1997, *J. Biotechnol.* **1997**, *57*, 83-90.
23. Conn, E.E. In β-*Glucosidases: Biochemistry and Molecular Biology*; Essen, A., Ed.; ACS Symposium Series 533; American Chemical Society: Washington, DC., 1993; pp. 15-26.
24. Glew, R.H., Gopalan, V.; Forsyth, G.W.; Van der Jagt, D.J. In β-*Glucosidases: Biochemistry and Molecular Biology*; Essen, A., Ed.; ACS Symposium Series 533; American Chemical Society: Washington, DC., 1993; pp. 82-112.
25. Poulton, J.E. *Plant Physiol.* **1990**, *94*, 401-405.
26. Kojima, M.; Poulton, J.E.; Thayer, S.S.; Conn, E.E. *Plant Physiol.* **1979**, *63*, 1022-1028.
27. Grabowski, G.A.; Berg-Fussman, A.; Grace, M. In β-*Glucosidases: Biochemistry and Molecular Biology*; Essen, A., Ed.; ACS Symposium Series 533; American Chemical Society: Washington, DC., 1993; pp. 66-81.
28. Nevalainen, H.; Penttila, M. In *The Mycota II, Genetics and Biotechnology*; Kück, U. Ed.; Springer-Verlag: Berlin, 1995, pp. 303-319.

164

29. Sternberg, D.; Vijayakumar, P.; Reese, E.T. *Can. J. Microbiol.* **1977**, *23*, 139-147.
30. Ladish, M.R.; Hong, J.; Voloch, M.; Tsao, G.T. *Basic Life Sci.* **1981**, *18*, 55-83.
31. Kadam, S.K.; Demain, A.L. *Biochem. Biophys. Res. Comm.* **1989**, *161*, 706-711.
32. Desrochers, M.; Jurasek, L.; Paige, M.G. *Appl. Environ. Microbiol.* **1981**, *23*, 222-228.
33. Blotkamp, P.J.; Takagi, M.; Pemberton, M.S.; Emert, G.H. *Am. Inst. Chem. Eng. Symp. Ser.* **1978** *74*, 85-90.
34. Savarese, J.J.; Young, S.D. *Biotechnol. Bioeng.* **1978**, *20*, 1291-1293.
35. Philippidis, G.P.; Smith, T.K.; Wyman, C.E. *Biotechnol. Bioeng.* **1993**, *41*, 846-853.
36. Dennis, C. *J. Gen. Microbiol.* **1972** *71*, 409-411.
37. Stevens, B.H.J.; Payne, J. *J. Gen. Microbiol.* **1977**, *100*, 381-393.
38. Barnett, J.A. *Adv. Carbohydr. Chem. Biochem.* **1976**, *32*, 125-234.
39. Gondé, P.; Blondin, B.; Ratamahenina, R.; Arnaud, A.; Galzy, P. *J. Ferment. Technol.* **1982**, *60*, 579-584.
40. Freer, S.N.; Detroy, R.W. *Biotechnol. Bioeng.* **1983**, *25*, 541-557.
41. Spindler, D.D.; Wyman, C.E.; Grohmann, K.; Philippidis, G.P. *Biotech. Lett.* **1992**, *14*, 403-407.
42. Freer, S.N.; Detroy, R.W. *Biotechnol. Lett.* **1982**, *4*, 453-458.
43. Gondé, P.; Blondin, B.; LeClere, M.; Ratomahenina, R.; Arnaud, A.; Galzy, P. *Appl. Environ. Microbiol.* **1984**, *48*, 265-269.
44. Freer, S.N. *Appl. Environ. Microbiol.* **1991**, *57*, 655-659.
45. Fleming, L.W.; Duerksen, J.D. *J. Bacteriol.* **1967**, *93*, 135-141.
46. Marchin, G.L.; Duerksen, J.D. *J. Bacteriol.* **1969**, *97*, 237-243.
47. Freer, S.N.; Greene, R.V. *J. Biol. Chem.* **1990**, *265*, 12864-12868.
48. Kilker, Jr., R.D.; Saunier, B.; Tkacz, J.S.; Herscovics, A. *J. Biol. Chem.* **1981**, *256*, 5299-5303.
49. Leclerc, M.; Arnaud, A.; Ratomahenina, R.; Galzy, P. Biotech. *Genetic Eng. Rev.* **1987**, *5*, 269-295.
50. Saha, B.C.; Bothast, R.J. *Biotech. Lett.* **1996**, *18*, 155-158.
51. Vasserot, Y.; Christiaens, H.; Chemardin, P.; Arnaud, A.; Galzy, P. *J. Appl. Bacteriol.* **1989**, *66*, 271-279.
52. Gueguen, Y.; Chemardin, P.; Arnaud, A.; Galzy, P. *Biotech. Appl. Biochem.* **1994**, *20*, 185-198.
53. Blondin, B.; Ratomanahena, R.; Arnaud, A,; Galzy, P. *Eur. J. Appl. Microbiol. Biotechnol.* **1983**, *17*, 1-6.
54. Gondé, P., Ratomanahena, R.; Arnaud, A.; Galzy, P. *Can. J. Biochem. Cell Biol.* **1985**, *63*, 1160-1166.
55. Drider, D.; Pommares, P.; Chemardin, P.; Arnaud, A.; Galzy, P. J. *Appl. Bacteriol.* **1993**, *74*, 473-479.

56. Gueguen, Y.; Chemardin, P.; Arnaud, A.; Galzy, P. *J. Appl. Bacteriol.* **1995**, *78*, 270-280.
57. Skory, C.D.; Freer, S.N.; Bothast, R.J. *Appl. Microbiol. Biotechnol.* **1996**, *46*, 353-359.
58. Freer, S.N. *Arch. Biochem. Biophys.* **1985**, *243*, 515-522.
59. Himmel, M.E.; Tucker, M.P.; Lastick, S.M.; Oh, K.K.; Fox, J.W.; Spindler, D.D.; Grohmann, K. *J. Biol. Chem.* **1986**, *261*, 12948-12955.
60. Freer, S.N. *J. Biol. Chem.* **1993**, *268*, 9337-9342.
61. Skory, C.D.; Freer, S.N. *Appl. Environ. Microbiol.* **1995**, *61*, 518-525.
62. Saha, B.C.; Bothast, R.J. *Appl. Environ. Microbiol.* **1996**, *62*, 3165-3170.
63. Vasserot, Y.; Chemardin, P.; Arnaud, A.; Galzy, P. *J. Basic Microbiol.* **1991**, *31*, 301-312.
64. Vasserot, Y.; Arnaud, A.; Galzy, P. *Bioresource Technol.* **1993**, *43*, 269-271.
65. Sánchez-Torres, P.; González-Candelas, L.; Ramón, D. *J. Agric. Food Chem.* **1998**, *46*, 354-360.
66. LeClere, M.; Gondé, P.; Arnaud, A.; Ratomahenina, R.; Galzy, P.; Nicolas, M. *J. Gen. Appl. Microbiol.* **1984**, *30*, 509-521.
67. Duerksen, J.D.; Halvorson, H. *J. Biol. Chem.* **1958**, *233*, 1113-1120.
68. Kempton, J.B.; Withers, S.G. *Biochemistry* **1992**, *31*, 9961-9969.
69. Trimbur, D.; Warren, R.A.J.; Withers, S.G. In β-*Glucosidases: Biochemistry and Molecular Biology*; Essen, A., Ed.; ACS Symposium Series 533; American Chemical Society: Washington, DC., 1993; pp. 42-55.
70. Legler, G. *Adv. Carbohydr. Chem. Biochem.* **1990**, *48*, 319-384.
71. Jabon, G.; Arnaud, A.; Galzy, P. *FEMS Microbiol. Lett.* **1994**, *118*, 207-212.
72. Janbon, G.; Derancourt, J.; Chemardin, P.; Arnaud, A.; Galzy, P. *Biosci. Biotech. Biochem.* **1995**, *59*, 1320-1322.
73. Woodward, J.; Wiseman, A. *Enzyme Microbiol. Technol.* **1983**, *4*, 73-79.
74. Johnston, M.; Carlson, M. In *The molecular and cellular biology of the yeast Saccharomyces*; Broach, J.; Jones, E.W.; Pringle, J., Ed.; Cold Spring Harbor Laboratory Press: Plainview, N.Y., 1993, Vol 2, pp. 193-281.
75. Entian, K-D.; Barnett, J.A. *Trends Biochem. Sci.* **1992**, *17*, 506-510.
76. Gancedo, J.M. *Eur. J. Biochem.* **1992**, *206*, 297-313.
77. Perlman, D.; Halvorson, H.O. *Cell* **1981**, *25*, 525-536.
78. Chu, F.K.; Maley, F. *J. Biol. Chem.* **1980**, *255*, 6392-6397.
79. Gascon, S.; Lampen, J.O. *J. Biol. Chem.* **1968**, *243*, 1573-1572.
80. deKroon R.A.; Koninsberger, W. *Biochim. Biophys. Acta.* **1970**, *204*, 590-609.
81. Duerksen, J.D.; Halvorson, H.O. *Biochim. Biophys. Acta* **1959**, *30*, 47-55.
82. MacQuillan, A.M.; Winderman, S.; Halvorson, H.O. *Biochem. Biophys. Res. Commun.* **1960**, *3*, 77-80.
83. MacQuillan, A.M.; Halvorson, H.O. *J. Bacteriol.* **1962**, *84*, 23-30.
84. Freer, S.N.; Detroy, R.W. *Appl. Environ. Microbiol.* **1985**, *50*, 152-159.
85. Pandey, M.; Mishra, S. *J. Ferment. Bioeng.* **1995**, *80*, 446-453.
86. Jabon, G.; Magnet, M.; Arnaud, A.; Galzy, P. *Gene* **1995**, *165*, 109-113.

87. Raynal, A.; Gueromeau, M. *Mol. Gen. Genet.* **1984**, *195*, 108-115.
88. Kohci, C.; Toh-e, A. *Gene* **1985**, *13*, 6273-6282.
89. Kohchi, C.; Toh-e, A. *Mol. Gen. Genet.* **1986**, *203*, 89-94.
90. Machida, M.; Ohtsuki, I.; Fukui, S.; Yamashita, I. *Appl. Environ. Microbiol.* **1988**, *54*, 3147-3155.
91. Henrissat, B. *Biochem. J.* **1991**, *280*, 309-316.
92. Henrissat, B.; Bairoch, A. *Biochem J.* **1993**, *293*, 781-788.
93. Leclerc, M.; Chemardin, P.; Arnaud, A.; Ratomahenina, R.; Galzy, P.; Gerbaud, C.; Raynal, A. *Arch. Microbiol.* **1986**, *146*, 115-117.
94. Skory, C.D.; Freer, S.N.; Bothast, R.J. *Curr. Genet.* **1996**, *30*, 417-422.
95. Leclerc, M.; Chemardin, P.; Arnaud, A.; Ratomahenina, R.; Galzy, P.; Gerbaud, C.; Raynal, A.; Guérineau, M. *Biotech. Appl. Biochem.* **1987**, *9*, 410-422.
96. Hu, A.S.L.; Epstein, R.; Halvorson, H.O.; Bock, R.M. *Arch. Biochem. Biophys.* **1960**, *91*, 210-219.
97. Kohchi, C.; Hayashi, M.; Nagai, S. *Agric. Biol. Chem.* **1985**, *49*, 779-784.
98. Villa, T.G.; Notario, V.; Villanueva, J.R. *FEMS Microbiol. Lett.* **1979**, *6*, 91-94.
99. Roth, W.W.; Srinivasan, V.R. *Prep. Biochem.* **1978**, *8*, 57-71.
100. Kilian, S.G.; Prior, B.A.; Venter, J.J.; Lategan, P.M. *Appl. Microbiol. Biotechnol.* **1985**, *21*, 148-153.
101. Herman, A.; Halvorson, H.O. *J. Bacteriol.* **1963**, *85*, 901-910.
102. Freer, S.N. *Can. J. Microbiol.* **1995**, *41*, 177-185.
103. Freer, S.N.; Skory, C.D. *Can. J. Microbiol.* **1996**, *42*, 431-436.

Chapter 11

Enzymology of Xylan Degradation

Badal C. Saha and Rodney J. Bothast

Fermentation Biochemistry Research Unit, National Center for Agricultural
Utilization Research, Agricultural Research Service, U.S. Department of Agriculture,
1815 N. University Street, Peoria, IL 61604

Xylan is a major component of the hemicellulose portion of plant
cell walls and constitutes up to 35% of the total dry weight of higher
plants. It consists of a homopolymeric backbone chain of β-1,4-
linked D-xylose units and short side chains including L-
arabinofuranosyl, O-acetyl, D-glucuronosyl or O-methyl-D-
glucuronosyl residues. Xylan degrading enzymes are produced by
a wide variety of aerobic and anaerobic fungi and bacteria.
Enzymatic hydrolysis of xylan involves a multi-enzyme system,
including endo-xylanase, β-xylosidase, α-arabinofuranosidase, α-
glucuronidase, acetylxylan esterase, ferulic acid esterase and p-
coumaric acid esterase. Synergistic interactions of all these enzymes
are required for the complete degradation of xylans. A brief review
of each enzyme involved in xylan degradation is presented.

Hemicelluloses, the second most common polysaccharides in nature, represent about
20-35% of lignocellulosic biomass (*1*). Xylans are the most abundant hemicelluloses.
In recent years, xylan degrading enzymes have received much attention because of their
practical applications in various agro-industrial processes, such as efficient conversion
of hemicellulosic biomass to fuels and chemicals, delignification of paper pulp,
digestibility enhancement of animal feedstock, clarification of juices and improvement
in the consistency of beer (*2-4*). Xylanases are of great interest to the paper and pulp
industry due to their bleach boosting properties (biobleaching of pulp), which reduces
environmentally unfriendly chlorine consumption (*5,6*). Cellulase-free xylanase can
facilitate lignin removal from paper pulp without any harmful effect. For this purpose,
the most advantageous working conditions for xylanases are high temperature and
alkaline pH. The utilization of hemicellulosic sugars is essential for efficient
conversion of lignocellulosic materials to ethanol. Dilute acid pretreatment at high
temperature converts hemicellulose to monomeric sugars (*7*). However, this
pretreatment also produces by-products that are toxic to fermentative microorganisms.

Other pretreatments, such as alkali, alkaline peroxide and ammonia fiber explosion (AFEX), solubilize xylans and produce xylooligosaccharides (8,9). These xylooligosaccharides are not fermentable and must be converted to monomeric sugars with enzymes or by other means prior to fermentation. Xylan degrading enzymes hold great promise in saccharifying various pretreated agricultural and forestry residues to monomeric sugars for fermentation to fuels and chemicals. Other potential applications of xylanases include biopulping of wood, coffee processing, fruit and vegetable maceration and preparation of high fiber baked goods (10). Xylose can be converted into xylitol, a potentially attractive sweetening agent, by catalytic reduction or by a variety of microorganisms (11). In addition, xylan degrading enzymes play a great role in elucidating the structures of complex xylans. In this article, we will give a brief review regarding the various enzymes involved in xylan degradation.

Structure of Xylan

Hemicelluloses are heterogeneous polymers of pentoses (D-xylose, L-arabinose), hexoses (D-mannose, D-glucose, D-galactose) and sugar acids. Unlike cellulose, hemicelluloses are not chemically homogeneous. Hardwood hemicelluloses contain mostly xylans, whereas softwood hemicelluloses contain mostly glucomannans (12). Xylans of many plant materials are heteropolysaccharides with homopolymeric backbone chains of 1,4-linked β-D-xylopyranose units. Besides D-xylose, xylans may contain L-arabinose, D-glucuronic acid or its 4-O-methyl ether, and acetic, ferulic, and p-coumaric acids. The frequency and composition of branches are dependent on the source of xylan (13). Xylans from different sources, such as grasses, cereals, softwood and hardwood, differ in composition. Birch wood (Roth) xylan contains 89.3 % D-xylose, 1% L-arabinose, 1.4% D-glucose and 8.3% anhydrouronic acid (14). Rice bran neutral xylan contains 46% D-xylose, 44.9% L-arabinose, 6.1% D-galactose, 1.9% D-glucose and 1.1% anhydrouronic acid (15). Wheat arabinoxylan contains 65.8% D-xylose, 33.5% L-arabinose, 0.1% D-mannose, 0.1% D-galactose and 0.3% D-glucose (16). Corn fiber hemicellulose contains 48-54% D-xylose, 33-35% L-arabinose, 5-11% D-galactose and 3-6% D-glucuronic acid (17). The schematic structure of the sugar moiety of the heteroxylans from corn fiber proposed by Saulnier et al. (18) based on a number of previous studies (18-23) is given in Figure 1. In softwood heteroxylans, L-arabinofuranosyl residues are esterified with p-coumaric acids and ferulic acids (24, 25). In hardwood xylans, 60-70% of the D-xylose residues are acetylated (26).

Biodegradation of Xylan

The total biodegradation of xylan requires endo-β-1,4-xylanase, β-xylosidase and several accessory enzymes, such as α-arabinofuranosidase, α-glucuronidase, acetylxylan esterase, ferulic acid esterase and p-coumaric acid esterase, which are necessary for hydrolyzing various substituted xylans. Table I lists the enzymes involved in the degradation of xylan and their modes of action. A hypothetical plant heteroxylan fragment and sites of attack by xylanolytic enzymes are shown in Figure 2.

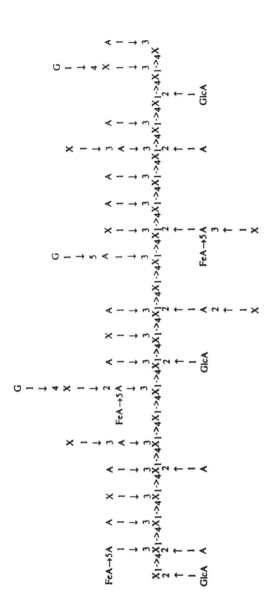

Figure 1. Schematic structure of heteroxylan from maize bran. (Reprinted from Saulnier, L.; Marot, C.; Chanliaud, E.; Thibault, J.-F. *Carbohydr. Polym.* **1995**, *26*, 279-287, with permission from Elsevier Science, Ref. *18*)

Table I. Major Enzymes Involved in the Biodegradation of Heteroxylans

Enzyme	Mode of Action
Endo-xylanase	Hydrolyzes mainly interior β-1,4- xylose linkages of the xylan backbone
Exo-xylanase	Hydrolyzes β-1,4- xylose linkages releasing xylobiose
β-Xylosidase	Releases xylose from xylobiose and short chain xylooligosaccharides
α-Arabinofuranosidase	Hydrolyzes terminal nonreducing α-arabinofuranose from arabinoxylans
α-Glucuronidase	Releases glucuronic acid from glucuronoxylans
Acetylxylan esterase	Hydrolyzes acetylester bonds in acetyl xylans
Ferulic acid esterase	Hydrolyzes feruloylester bonds in xylans
p-Coumaric acid esterase	Hydrolyzes p-coumaryl ester bonds in xylans

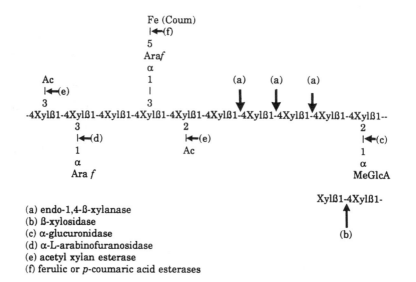

(a) endo-1,4-β-xylanase
(b) β-xylosidase
(c) α-glucuronidase
(d) α-L-arabinofuranosidase
(e) acetyl xylan esterase
(f) ferulic or p-coumaric acid esterases

Figure 2. Representative plant xylan and the sites of cleavage by xylan-degrading enzymes. Ac, acetyl group; Araf, L-arabinofuranosidase, MeGlcA, 4-O-methyl-D-glucuronic acid; Xyl, D-xylose; Fe, ferulic acid; Coum, p-coumaric acid. (Reprinted from Wood, T. M.; Castanares, A.; Smith, D. C.; McCrae, S. I.; Brown, J. A. In *Xylans and Xylanases*; Visser, J.; Beldman, G.; Kusters-Van Someren, M. A.; Voragen, A. G. J., Eds.; Elsevier Sciences Publishers, Amsterdam, **1992**; pp.187-202, with permission from Elsevier Science, Ref. *27*)

The endo-xylanase attacks the main chains of xylans and β-xylosidase hydrolyzes xylooligosaccharides to xylose. The α-arabinofuranosidase and α-glucuronidase remove the arabinose and 4-*O*-methyl glucuronic acid substituents, respectively, from the xylan backbone. The esterases hydrolyze the ester linkages between xylose units of the xylan and acetic acid (acetylxylan esterase) or between arabinose side chain residues and phenolic acids, such as ferulic acid (ferulic acid esterase) and *p*-coumaric acid (*p*-coumaric acid esterase). Synergistic action between depolymerizing and side-group cleaving enzymes has been verified using acetylated xylan as a substrate (*28*). Bachmann and McCarthy (*29*) reported significant synergistic interaction among endo-xylanase, β-xylosidase, α-arabinofuranosidase and acetylxylan esterase of the thermophilic actinomycete *Thermomonospora fusca*. Many xylanases do not cleave glycosidic bonds between xylose units which are substituted. The side chains must be cleaved before the xylan backbone can be completely hydrolyzed (*30*). On the other hand, several accessory enzymes only remove side chains from xylooligosaccharides. These enzymes require a partial hydrolysis of xylan before the side chains can be cleaved (*31*). Although the structure of xylan is more complex than cellulose and requires several different enzymes with different specificities for complete hydrolysis, the polysaccharide does not form tightly packed crystalline structures like cellulose and is, thus, more accessible to enzymatic hydrolysis (*32*). Many microorganisms, such as *Penicillium capsulatum* and *Talaromyces emersonii*, possess complete xylan degrading enzyme systems (*33*). Many rumen bacteria possess the capacity to utilize a variety of xylooligosaccharides (*34-36*).

Endo-xylanase

Endo-xylanase (β-1,4-D-xylan xylanohydrolase, EC 3.2.1.8) is the key enzyme for xylan depolymerization. Xylanases have been widely detected in bacteria and fungi and many microorganisms produce multiple forms of the enzyme. Comparative properties of some fungal and bacterial xylanases are presented in Tables II and III, respectively. The yeast-like fungus *Aureobasidium* is a good source of xylanase (molecular weight, MW, 20,000) with an exceptionally high specific activity (2,100 U/mg protein) (*37*). Xylanase represents nearly half of the total extracellular protein and yields of up to 0.3 g of xylanase per liter can be obtained (*38*). Berenger *et al.* (*39*) purified three xylanases (A, B and C), ranging in size from MW 44,000 to 72,000, from *Clostridium stercorarium*. Esteban *et al.* (*40*) reported that *Bacillus circulans* produces two xylanases, a high MW enzyme with a pI of 4.5 and a low MW enzyme with a pI of 9.0. *Bacillus polymyxa* produces a basic xylanase with a pI of 9.0 and an acidic xylanase with a pI of 4.9 (*41*). Comtat (*42*) separated nine xylanases from *Sporotrichum dimorphosporum* by DEAE-Sephadex and preparative isoelectric focusing. *Aeromonas caviae* W-61 produces three xylanases with MW's of 22,000, 41,000 and 58,000, pI values of 9.2, 11.5 and 2.5 and different temperature and pH optima (*43,44*). Fujimoto *et al.* (*45*) purified three distinct xylanases (FIa, FIb and FIII) from *Aspergillus aculeatus*. The MW's of FIa, FIb and FIII were 18,000, 26,000, and 52,000 and their pI values were pH 5.6, 9.0, and 3.8, respectively. The xylanases from extremely

Table II. Comparative Properties of Some Fungal Endo-xylanases

Organism		Molecular Weight	pI	Optimum Temperature (°C)	Optimum pH	Hydrolysis Products[a]
Aspergillus awamori (52)	I	39,000	5.7-6.7	55	5.5-6.0	X_1, X_2, X_3
	II	23,000	3.7	50	5.0	X_2, X_3, A
	III	26,000	3.3-3.5	45-50	4.0	X_2, X_3
Aureobasidium pullulans (37)		20,000	8.5	45	4.5	-
Aureobasidium pullulans (53)		25,000	9.4	54	4.8	-
Cryptococcus flavus (54)		25,000	10.0	55	4.5	X_2, X_3, X_4, X_n
Fusarium oxysporum (55)		80,000	-	50	5.0	X_n
Humicola grisea var. thermoidea (56)		25,500	-	70	5.5	X_n
Neocallimastrix frontalis (57)	I	45,000	-	55	6.0	X_n
	II	70,000		55	5.5	X_n
Neurospora crassa (58)	I	33,000	4.5	50	4.8	X_1 to X_5, A
	II	30,000	4.5	50	4.8	X_1 to X_5, A
Penicillium chrysogenum (59)		35,000	4.2	40	6.0	X_1, X_2
Talaromyces emersonii (60)	II	74,800	5.3	78	4.2	X_2, X_3, X_4, X_n
	III	54,200	3.8	67	3.5	X_2, X_3, X_4, X_n
Trichoderma harzianum (61-63)	20kDA	20,000	9.4	50	5.0	X_2, X_n
	22kDA	22,000	8.5	45-50	4.5-5.0	-
	29kDA	29,000	9.5	60	5.0	X_2, X_n
Trichoderma koningii (64)	1	29,000	7.2	60	4.9-5.8	X_1, X_2, X_3, A
	2	18,000	7.3	50	4.9-5.5	X_2, X_3, X_4
Trichoderma reesei (65)	I	19,000	5.5	-	4.0-4.5	X_1, X_2, X_3, X_n
	II	20,000	9.0	-		
Trichoderma viride (66)		22,000	9.3	53	5.0	X_1, X_2

[a] X_1, xylose; X_2, xylobiose; X_3, xylotriose; X_4, xylotetraose; X_5, xylopentaose; X_n, xylooligosaccharides; A, L-arabinose.

Table III. Comparative Properties of Some Bacterial Endo-xylanases

Organism		Molecular Weight	pI	Optimum Temperature (°C)	Optimum pH	Hydrolysis Products[a]
Bacillus pumilus (67)		24,000	-	45-60	6.5	X_2, X_3, X_4, X_n
Bacillus stearothermophilus (48)		43,000	7.0	75	6.5	X_1, X_2
Bacillus subtilis (68)		32,000	-	50	5.0	X_1, X_2, X_3
Clostridium acetobutylicum (69)	A	65,000	4.4	50	5.0	X_2, X_3, X_4, X_5
	B	29,000	8.5	60	5.5-6.0	X_2, X_3
Clostridium stercorarium (39)	A	44,000	4.5	75	5.5-7.0	X_1, X_2, X_3, X_4
	B	72,000	4.4	75	5.5-7.0	X_1, X_2, X_3, X_4
	C	62,000	4.3	75	5.5-7.0	X_1, X_2, X_3, X_4
Fibrobacter succinogenes (70)	1	53,700	8.9	39	7.0	$X_1, X_2, X_3, X_4, X_n, A$
	2	66,000	8.0	55	6.3	X_1, X_2, X_3, X_4
Streptomyces viridosporus (71)		59,000	10.2-10.5	65-70	7.0-8.0	-
Streptomyces thermoviolaceus (72)	I	54,000	4.2	70	7.0	X_1, X_2
	II	33,000	8.0	60	7.0	X_1, X_2, X_3
Thermomonospora curvata (73)	1	36,000	4.2	75	7.8	X_2, X_3, X_n
	2	19,000	7.1	75	7.2	X_3, X_4, X_n
	3	15,000	8.4	75	6.8	X_3, X_4, X_5, X_n
Thermotoga sp. (46)		31,000	-	105	5.3	X_2, X_3
Thermotoga maritima (46)	A	120,000	-	92	6.2	X_2, X_3
	B	40,000	5.6	105	5.4	X_2, X_3
Thermotoga thermarum (74)	1	266,000	-	80	6.0	X_2, X_4, X_5, X_n
	2	35,000	-	90-100	7.0	X_2

[a] X_1, xylose; X_2, xylobiose; X_3, xylotriose; X_4, xylotetraose; X_5, xylopentaose; X_n, xylooligosaccharides; A, L-arabinose.

thermophilic *Thermotoga* sp. and *Thermotoga maritima* had an optimum temperature of 105°C and are the most thermostable xylanases reported to date (*46,47*). *Bacillus stearothermophilus* T-6 produced an extracellular xylanase that bleached pulp optimally at pH 9.0 and 65°C (*48*). The xylanases from thermophilic *Bacillus thermoleovorans* strain K-3d and *Bacillus flavothermus* strain LB3A were optimally active at 70-80°C and 70 °C, respectively, and pH 7.0 (*49*). Overall, bacterial xylanases tend to exhibit higher optimal temperatures and pH's than fungi and produce a variety of hydrolysis products. One xylanase from *Aeromonas caviae* ME-1 produced xylobiose exclusively and another one only xylotriose from xylan (*50,51*). A few recent reviews have dealt with the multiplicity, structure and function of microbial xylanases, and molecular biology of xylan degradation (*2,75,76*).

Xylanases hydrolyze glycosidic bonds via a double displacement, general acid-base mechanism (*77*). The primary sequences of over 70 xylanases have been determined (*78*). Xylanases have been classified into several families on the basis of sequence similarities (*79*). Wong *et al.* (*2*) divided xylanases into two categories : (1) low MW (16,000 to 22,000), basic (pI 8.3-10.0) xylanases and (2) high MW (43,000 to 50,000), acidic (pI 3.6 to 4.5) xylanases. These groups roughly correspond to families F/10 and G/11 of the glycosyl hydrolase classification (*80*). Xylanases from both families hydrolyze the β-1,4-glycosidic bond of xylan with the retention of the anomeric configuration (*81*). Several xylanases from a variety of organisms contain cellulose binding domains (CBD's) (*82-84*). It has been suggested that cellulose acts as a general receptor for plant cell wall hydrolases as it is the only structural polysaccharide that does not vary in structure between plant species, apart from small changes in the degree of crystallinity (*32*). Recently, Sun *et al.* (*85*) showed that xylanase A from *Clostridium stercorarium* binds to insoluble oat-spelt xylan via its CBD's and the CBD's play an important role in the hydrolysis of insoluble xylan. In *Ruminococcus flavefaciens*, a single gene, *xynD,* encodes a bifunctional enzyme having separate xylanase and β- (1,3-1,4)-glucanase domains (*86*).

The structures of the catalytic domains of endo-xylanases (family G/11) from *Bacillus pumilus* (*87*), *Bacillus circulans* (*88*) and *Trichoderma harzianum* (*88*) have been determined. Torronen and Rouvinen (*89*) determined the three-dimensional structures of two major xylanases, XYNI and XYNII, from *Trichoderma reesei*. They reported that the amino acid sequences of both enzymes are highly homologous (identity ~50%) and both XYNI and XYNII exist as a single domain that contains two mostly antiparallel β-sheets which are packed against each other. The β-sheet structure is twisted, forming a cleft where the active site is situated. Two glutamic acids in the cleft (Glu 75 and Glu 164 in XYNI and Glu 86 and Glu 177 in XYNII) are most likely involved in catalysis. Figure 3 shows the overall structures of XYNI (panel a) and XYNII (panel b) [drawn using MOLSCRIPT; Kraulis, 1991 (*90)*] and the analogy with the human right hand (panel c) (*91*). In the case of xylanase I (family G) from *Aspergillus niger*, two conserved glutamate residues, Glu 79 and Glu 170, are likely involved in catalysis (*92*). Similar findings were reported in other members of the xylanase family G (*93-96*).

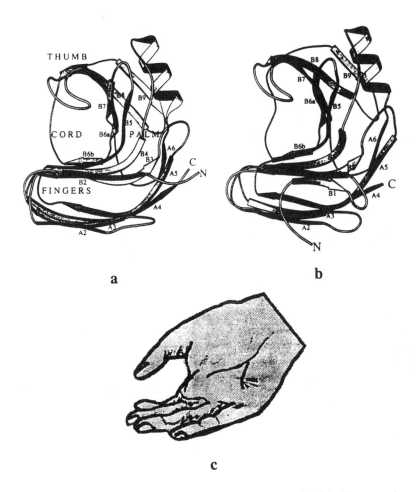

a

b

c

Figure 3. Overall structures of XYNI (a) and XYNII (b) (drawn using MOLSCRIPT; Kraulis, 1991 [80]) from *Trichoderma reesei*. The analogy with the human right hand (c) is shown. (Reprinted from Torronen, A.; Rouvinen, J. *Biochemistry* **1995**, *34*, 847-856, with permission from American Chemical Society, Washington, D. C. Ref. *89*).

β-Xylosidase

β-Xylosidase (1,4-β-D-xylan xylanohydrolase, EC 3.2.1.37), which hydrolyzes xylobiose and short xylooligosaccharides from the nonreducing end to xylose, is essential for complete breakdown of xylans. The enzyme is part of most microbial xylanolytic systems. The β-xylosidase is cell-associated in most bacteria and yeast (*91*), but it is found free in culture media of some fungi (*97*). Comparative properties of some β-xylosidases are given in Table IV. In comparison with endo-xylanases, β-xylosidases have larger molecular weights (between 60,000 and 360,000) and often have two or more subunits (*10*). In *Bacteroides ovatus*, the β-xylosidase, rather than xylanase, may be rate-limiting in xylan degradation (*98*). The enzyme from *Neocallimastix frontalis* is a dimer (native MW of 150,000) of two different subunits with MW's of 83,000 and 53,000 (*99*). β-Xylosidases with two different subunits have been reported in *Trichoderma lignorum* (*100*), *Chaetomium trilaterale* (*101*) and *Clostridium acetobutylicum* (*102*). The β-xylosidase from *Thermomonospora fusca* has a MW of 168,000 with three subunits (*103*).

The rate of hydrolysis of xylooligosaccharides by β-xylosidases usually decreases with increasing chain length (*106,114,119,120*). In contrast, the β-xylosidase from *Trichoderma lignorum* releases xylose residues from xylohexaose, xyloheptaose and xylooctaose more rapidly than from shorter chain oligosaccharides and xylobiose (*100*). The β-xylosidase from *Trichoderma reesei* RUT C-30 is a multi-functional enzyme capable of hydrolyzing xylooligosaccharides (DP 2-7). The apparent V_{max} with this enzyme increases with increasing chain length (*121*). It also attacks debranched beech wood xylan and methylglucuronoxylan, forming xylose as the only end product and exhibits α-arabinofuranosidase activity. In addition to releasing xylose, many β-xylosidases generate transfer products of higher MW than that of the substrate (*122-125*). Yasui *et al.* (*124*) reported the production of nonreducing xylobiose from xylobiose by transxylosylation with β-xylosidase from *Aspergillus niger*. Sulistyo *et al.* (*126*) reported the synthesis of hydroquinone β-xyloside from xylooligosaccharides to acceptor hydroquinone by transxylosylation with crude β-xylosidase from *Aspergillus pulverulentus*. The two β-xylosidases from this organism have broad acceptor specificities in transferring the xylosyl residue of xylooligosaccharides to various alcohols and phenolic compound acceptors (*127*). Most β-xylosidases are competitively inhibited by xylose (*99,118*).

β-Xylosidases, based on amino acid sequence similarities, have been grouped into families 39 and 43 of a general glycosyl hydrolase classification (*80,128*). The stereochemistry of substrate hydrolysis by these two families of β-xylosidases has been investigated (*129,130*). Braun *et al.* (*129*) showed that the recombinant β-xylosidase from *Butyrivibro fibrisolvens* belongs to family 43 and operates by a molecular mechanism that results in anomeric inversion. Armand *et al.* (*130*) reported that the opposite molecular mechanism (anomeric retention) prevails for the β-xylosidase from *Thermoanaerobacter saccharolyticum*, a member of family 39. This enzyme also possesses a high transglycosylating activity and produces a variety of β-linked

Table IV. Comparative Properties of Some Microbial β-Xylosidases

Organism		Molecular Weight	pI	Optimum Temperature (°C)	Optimum pH	Inhibition by xylose (Ki, mM)
Arxula adeninivorans (104)	bX1	120,000	-	60	5.0	5.8
	bX2	60,000	-	60	5.0	5.8
Aspergillus nidulans (105)		180,000	3.4	50	5.0	Inhibited by xylose
Aureobasidium pullulans (106)		224,000	-	80	4.5	Inhibited by xylose
Bacillus stearothermophillus (107)		150,000	4.2	70	-	-
Clostridium acetobutylicum (102)		224,000	5.85	45	6.0-6.5	-
Clostridium cellulolyticum (108)		43,000	-	35	7.5	
Humicola grisea var. thermoidea (109)		43,000	4.0	50	6.0	-
Neomallimastrix frontalis (99)		150,000	4.6	37	6.4	3.98
Neomallimastrix patriciarum (110)	I	39,500	4.7	50	6.0	-
	II	150,000	4.7	40	6.0	-
Penicillium wortmanni (111)	1	110,000	3.7	3-4.5	55-65	-
	2	195,000	4.28	3-4.5	55-65	-
	3	210,000	4.6	3-4.5	55-65	-
	4	180,000	4.8	3-4.5	55-65	-
Rhodothermus marinus (112)		169,000	-	90	6.0	-
Sclerotium rolfsii (113)		170,000	6.8	50	4.5	-
Thermoanaerobacter ethanolicus (114)		165,000	4.6	93	5.8-6.0	-
Thermomonospora curvata (115)		112,000	4.8	60-68	6-7	42.5
Thermomonospora fusca (116)		168,000	4.37	-	-	19
Trichoderma harzianum (117)		60,000	-	70	4.0-4.5	Inhibited by xylose
Trichoderma reesei (118)		100,000	4.7	60	4	2.3

xylooligosaccharides. The β-xylosidase from *Clostridium cellulolyticum* was found to act by inverting the β-anomeric configuration (*108*).

α-Arabinofuranosidase

α-Arabinofuranosidases (α-L-arabinofuranoside arabinofuranohydrolase, EC 3.2.1.55) are exo-type enzymes, which hydrolyze terminal nonreducing residues from arabinose containing polysaccharides. These enzymes can hydrolyze (1→3)- and (1→5)-α-arabinosyl linkages of arabinan. The α-arabinofuranosidases are part of microbial xylanolytic systems necessary for complete breakdown of arabinoxylans (*29,30,131,132*). Comparative properties of some microbial α-arabinofuranosidases are summarized in Table V. The authors suggest α-arabinofuranosidases warrant substantial research efforts because they represent potential rate limiting enzymes in xylan degradation, particularly those substrates from agricultural residues (fiber, stover, etc.).

Multiple forms of α-arabinofuranosidases have been detected in the culture broth of *Aspergillus nidulans* (*133*), *Aspergillus niger* (*134*), *Aspergillus terreus* (*135*) and *Penicillium capsulatum* (*136*). The α-arabinofuranosidase from *Aureobasidium pullulans* is a homodimer with an apparent native MW of 210,000 and a subunit MW of 105,000 (*137*). Komae *et al.* (*138*) showed that the MW of α-arabinofuranosidase from *Streptomyces purpurascens* IFO 3389 is about 495,000 and the enzyme contains 8 equal subunits of MW 65,000. The MW of α-arabinofuranosidase of *Butyrivibrio fibrisolvens* GS113 is 240,000 and it consists of 8 subunits of MW 31,000 (*139*). The enzyme from *Clostridium acetobutylicum* ATCC 824 is a single polypeptide with a MW of 94,000 (*30*). The intracellular α-arabinofuranosidase from *Aspergillus niger* is a monomer with a MW of 67,000 (*140*). Microbial α-arabinofuranosidases have a broad range of pH and temperature dependence, with optimal activities occurring between pH 3.0- 6.9 and 40-75°C (*30,136,137,141-143*). The purified enzyme from *Rhodotorula flava* is highly acid stable, retaining 82% of its activity after being maintained for 24 h at pH 1.5 and at 30°C (*144*). Optimum activity of this enzyme is at pH 2.0.

The α-arabinofuranosidases from *Aureobasidium pullulans* (*137*), *Streptomyces* sp. strain 17-1 (*145*), *Streptomyces diastaticus* (*146*) and *Bacillus subtilis* 3-6 (*138*) have hydrolytic activity for both α-(1→3) and α-(1→5)-linked, non-reducing, terminal residues. They do not act on internal α-arabinosyl linkages. The α-arabinofuranosidases from *Aspergillus niger* (*147*) and *Streptomyces purpurascens* IFO 3389 (*135*) hydrolyze either (1→5) or (1→3)-arabinosyl linkages of arabinan. The enzyme purified from *Streptomyces purpurascens* is inactive against arabinans and arabinogalactans (*138*). Kormelink et al. (*148*) described another type of α-arabinofuranosidase that is active only on arabinoxylans. Recently, Van Laere *et al.* (*149*) described a new arabinofuranohydrolase from *Bifidobacterium adolescentis* able to remove arabinosyl residues from double-substituted xylose units in arabinoxylan. The arabinofuranosidase from *Streptomyces lividans* exhibits no activity against oat spelt xylan or arabinogalactan (*150*). It slowly acts on arabinan and arabinoxylans by

Table V. Comparative Properties of Some Microbial α-Arabinofuranosidases

Organism		Molecular Weight	pI	Optimum Temperature (°C)	Optimum pH	Polymer Attacked[a]
Aspergillus awamori (145)		32,000	-	50	5.0	AX
Aspergillus nidulans (133)		65,000	3.3	65	4.0	BA, AX
Aspergillus niger 5-16						
Intracellular (151)		67,000	3.5	60	4.0	BA
Extracellular (147)		53,000	3.6	-	4.0	BA, AX
Aspergillus terreus (135)	I	39,000	7.5	-	3.5-4.5	OSX, RAX, AGX, BA
	II	59,000	8.3	-	3.5-4.5	OSX, RAX, AGX, BA
	III	59,000	8.5	-	3.5-4.5	OSX, RAX, AGX, BA
Aureobasidium pullulans (137)		210,000	-	75	4.0-4.5	BA, AX, OX
Bacillus stearothermophilus (141)		110,000	-	70	7.0	Used for delignification
Bacillus subtilis (154)		65,000	5.3		6.5	BA
Butyrivibrio fibrisolvens (139)		240,000	6.0	55	6.0-6.5	BA, AX, OSX
Clostridium acetobutylicum (30)		94,000	8.2		5.0-5.5	BA (inactive on OSX, AX)
Cytophaga xylanolytica (155)		160,000 -240,00	6.1	45	5.8	AX, BA
Penicillium capsulatum (136)	I	64,500	4.15	60	4.0	BA, AX, AXO
	II	62,700	4.54	55	4.0	BA, AX, AXO
Ruminococcus albus (131)		310,000	3.8		6.9	AH
Streptomyces sp. 17-1 (145)		92,000	4.4		6.0	BA, AX, AG
Streptomyces purpurascens (138)		495,000	3.9		6.5	A₂, A₃
Trichoderma reesei (132)		53,000	7.5		4.0	AX, AXO

[a] AX, arabinoxylan; BA, beet arabinan; CX, corn endosperm xylan; OX, oatspelt xylan: AXO, arabinoxylan oligosaccharides;A₂, arabinobiose; A₃, arabinotriose.

releasing arabinose after prolonged incubation. The limit of hydrolysis of arabinan by the α-arabinofuranosidase from *B. subtilis* 3-6 was only 15%, even when the enzyme was present in excess (*151*). A bifunctional protein with β-xylosidase and α-arabinofuranosidase activities from *Butyrivibrio fibrisolvens* has been reported (*152*). In *Bacteroides ovatus*, α-arabinofuranosidase and β-xylosidase activities were suggested to be catalyzed by a bifunctional protein or two proteins of very similar MW (*98,153*). One α-arabinofuranosidase from *Penicillium capsulatum* was competitively inhibited by arabinose with a K_i of 16.4 mM (*136*). Arabinose at 80 mM concentration caused a 40% reduction of the hydrolytic activity of α-arabinofuranosidase from *Aspergillus nidulans* (*142*). The enzyme from *Aureobasidium pullulans* was not inhibited by 1.2 M (21.6%) arabinose (*137*).

α-Glucuronidase

α-Glucuronidase (α-D-glucuronidase, EC 3.2.1.131) cleaves the α-1,2-glycosidic linkage between xylose and glucuronic acid or its 4-O-methyl ether. Hardwood xylans possess on average of one α-1,2 -linked uronic acid side group per ten xylose units and softwood xylans contain one per five xylose units (*26*). The production of α-glucuronidase activity has been reported in many fungi, such as *Agaricus bisporus* (*156*), *Pleurotus ostreatus* (*156*), *Aspergillus niger* (*157*), *Schizophyllum commune* (*157*), *Dactylium dendroides* (*158*), *Trichoderma* sp. (*159*), *Trichoderma reesei* (*160, 161*), *Thermoascus aurantiacus* (*162*) and *Tyromyces palustris* (*159*), and bacteria, such as *Fibrobacter succinogenes* (*163*), *Streptomyces* spp. (*164*), *Streptomyces flavogriseus* (*165*), *Streptomyces olivochromogenes* (*158*) and *Thermoanaerobactium* sp. (*166*). Siika-aho *et al.* (*161*) purified an α-glucuronidase from *Trichoderma reesei* RUT C-30. The enzyme had a MW of 91,000 and a pI of 5.0-5.2. Highest activity was observed on low MW xylooligomer substrates. The enzyme seems to act almost exclusively on the bond between the terminal xylose at the nonreducing end of a xylose chain and the methylglucuronic acid attached to it. In addition, the α-glucuronidase enhanced the hydrolysis of glucuronoxylan by pure xylanases. The purified α-glucuronidase from *Thermoascus aurantiacus* had a MW of 118,000 and acted optimally at pH 4.5 and 65°C (*162*). The enzyme specifically cleaves α-1,2-glycosidic linkage between 4-O-methyl-α-glucuronic acid and xylose in xylan and several glucurono-xylooligosaccharides. The purified α-glucuronidase from *Thermoanaerobacter* sp. had a MW of 130,000 (two subunits of MW 74,000) and a pI of 4.65 (*166*). It exhibited optimal activity at pH 5.4 and 60°C. It was very active on 4-O-methylglucuronosyl xylobiose, 4-O-methylglucuronosyl xylotriose and 4-O-methylglucuronosyl xylotetraose, but exhibited a low activity against 4-O-methylglucuronosyl xylan. The enzyme from *Agaricus bisporus* cleaved 4-O-methyl-α-D-glucuronic acid from substituted xylooligomers, but not from substituted xylan (*167*). The α-glucuronidases from *Aspergillus niger* and *Schizophyllum commune* exhibited activities against 4-O-methylglucuronic acid substituted xylans (*157*). The properties of selected α-glucuronidases are summarized in Table VI.

Table VI. Comparative Properties of Some Microbial α-Glucuronidases

Organism	Molecular Weight	pI	Optimum Temperature (°C)	Optimum pH	Preferred Substrate
Agaricus bisporicus (161)	450,000	2.6-2.9	-	3.3-3.8	Xylooligomers (no activity on xylan)
Thermomoanaerobacter sp. *(166)*	130,000	4.65	60	5.4	Xylooligomers
Thermoascus aurantiacus (162)	118,000	-	65	4.5	Xylooligomers, xylan
Trichoderma reesei (168)	91,000	5.0-5.2	60	4.5-5.0	Xylooligomers
Trichoderma reesei (169)	100,000	-	45	5.0	-

Acetylxylan Esterase

Acetylxylan esterases (EC 3.1.1.6) remove O-acetyl groups from the C-2 and C-3 positions of xylose residues in both xylan and xylooligosaccharides. Table VII shows the characteristics of representative acetylxylan esterases. Biely *et al.* (*170*) first reported the presence of acetylxylan esterases in various fungal cellulolytic and hemicellulolytic systems, such as *Trichoderma reesei*, *Aspergillus niger*, *Schizophyllum commune* and *Aureobasidium pullulans*. Acetylxylan esterase activity has been reported to be present in a number microorganisms, such as *Aspergillus awamori* (*158*), *Aspergillus japonicus* (*171*), *Aspergillus versicolor* (*171*), *Aspergillus oryzae* (*172*), *Fusarium oxysporum* (*173*), *Penicillium purpurogenum* (*174*), *Rhodotorula mucilaginosa* (*175*), *Schizophyllum commune* (*176, 177*), *Trichoderrma viride* (*178*), *Bacillus pumilus* (*179*), *Pseudomonas fluorescens* (*180*), *Butyrivibrio fibrisolvens* (*181*), *Streptomyces* spp. (*164*), *Streptomyces lividans* (*182*), *Thermoanaerobacter saccharolyticum* (*183*), *Thermoanaerobacterium* sp. (*184*), *Caldocellum saccharolyticum* (*185*), *Thermomonospora fusca* (*186*) and *Fibrobacter succinogenes* (*187*). A number of them have been purified and characterized. Acetylxylan esterase (AXE I) from *Trichoderma reesei* contains an active-site serine residue and a cellulose-binding domain (*188*). The acetylxylan esterase (AXE II) of *Penicillium purpurogenum* lacks a cellulose binding domain (*189*).

Acetylation of the xylose residues in xylan can impede forage cell wall digestibility *in vivo* and inhibit enzymatic degradation *in vitro* (*193,194*). The role of acetylxylan esterase is to create unsubstituted xylose residues, allowing endo-xylanase or β-xylosidase to hydrolyze glycosidic linkages (*195,196*). The activity of acetylxylan esterase has been reported to decrease with an increase in the degree of acetylation of the acetylxylan substrate (*196*). The purified acetylxylan esterase from *Schizophyllum commune* preferentially deacetylated the 3-position of methyl 2,3,4-tri-O-acetyl-β-xylopyranosides and 2,3,4,6-tetra-O-acetyl-β-glucopyranosides (*197*). Removal of the 3-acetyl group from the xylopyranoside was accompanied by a slower deacetylation at positions 2 and 4. The acetylxylan esterase from *Streptomyces lividans* rapidly deacetylated monoacetylated xylopyranosyl residues but attacked doubly acetylated residues slowly (*198*). The enzyme from *Trichoderma reesei* deacetylated monoacetylated xylopyranosyl residures more readily than di-O-acetylated residues of partially acetylated β-1,4-linked xylans (*199*). The two acetyl esterases from *Trichoderma reesei* showed very different substrate specificities. One had a preference for high MW substrates and the other showed high activity only toward acetylxylobiose (*200*). The latter enzyme was regioselective, cleaving only the acetyl substituents from the C-3 position of the xylose ring. The highest xylose yield from acetylated xylan was obtained by the synergistic action of xylanase, β-xylosidase and acetylxylan esterase.

Ferulic Acid and *p*-Coumaric Acid Esterases

Ferulic acid and *p*-coumaric acid are common constituents of animal forage feed and may represent up to 2.5% by weight of the cell walls in temperate grasses (*201*). Many

Table VII. Comparative Properties of Some Microbial Acetylxylan Esterases

Organism		Molecular Weight	pI	Optimum Temperature (°C)	Optimum pH
Aspergillus niger (190)		30,480	3.0-3.2	50	5.5-6.0
Bacillus pumilus (179)		190,000	4.8	55	8.0
Fibrobacter succinogenes (187)		55,000	4.0	45	7.0
Penicillium purpurogenum (174)	AXE 1	48,000	7.5	50	5.3
	AXE II	23,000	7.8	60	6.0
Schizophyllum commune (191)		31,000	-	30-45	7.7
Streptomyces lividans (179)		34,000	9.0	70	7.5
Thermoanaerobacterium sp. (184)	I	195,000	4.2	80	7.0
	II	106,000	4.3	84	7.5
Trichoderma reesei (192)	AXE I	34,000, 20,000	7.0	60-65	5.0-6.0
	AXE II	34,000, 20,000	6.8	60-65	5.0-6.0

of the arabinose residues in various arabinoxylans are esterified with ferulic and p-coumaric acid residues (202). Barley straw arabinoxylan contains approximately one p-coumaric acid per thirty-one arabinose residues and one ferulic acid per fifteen arabinose residues. As their generic names imply, ferulic acid esterases cleave ester linkages between ferulic acids and arabinose and p-coumaric acid esterases cleave ester linkages between p-coumaric acids and arabinose in xylans. MacKenzie *et al.* (165) first described ferulic acid esterase activity in hemicellulase preparations and suggested that phenolic acid esterases, such as ferulic acid esterase and p-coumaric acid esterase perform a function similar to alkali in the deesterification of plant tissues. Ferulic acid esterase activity has been reported in culture supernatants of a number of microorganisms, such as *Streptomyces olivochromogenes* (164), *Schizophyllum commune* (176), *Aspergillus niger* (203), *Aspergillus terreus* (204), *Trichoderma reesei* (157), *Fibrobacter succinogenes* (205) and *Orpinomyces* sp. (206). p-Coumaric acid esterase activity has been reported in the culture supernatants of *Streptomyces viridosporus* (207, 208), *Neomallimastix* (206), *Orpinomyces* sp. (206) and *Penicillum pinophilum* (209). Only a few of these esterases have been purified and characterized (210). Some characteristics of microbial ferulic acid and p-coumaric acid esterases are given in Table VIII. Recently, Bartolome *et al.* (211) showed that esterase from *Aspergillus niger* (Ferulic acid esterase III) and a recombinant *Pseudomonas fluorescens* subsp. *cellulosa* esterase (XylD) released 5-5'-ferulic dehydrodimer (diferulic acid) from barley and wheat cell walls.

Concluding Remarks

In recent years, substantial progress has been made in understanding the structure and function of endo-xylanases. At present, our particular goal at USDA for the application of xylanolytic enzymes is the complete hydrolysis of various pretreated hemicellulosic agricultural residues, such as corn fiber, to fermentable sugars. Pretreatment of any lignocellulosic material is crucial prior to enzymatic hydrolysis. Our results with hot water and alkali pretreated corn fiber demonstrate that corn fiber arabinoxylan is very resistant to enzymatic hydrolysis (unpublished data). Currently, there is no suitable commercial enzyme preparation that can efficiently hydrolyze pretreated corn fiber xylan to monomeric sugars. Xylooligosaccharides represent about 30-40% of the xylan degradation products and very little xylose is produced after enzymatic treatment of, for example, AFEX pretreated corn fiber (8,9). This research indicates the need for better xylan degrading enzymes (7). Ineffectiveness of commercial hemicellulases in degrading corn fiber xylans, as well as production of substances inhibitory to subsequent microbial fermentation during dilute acid pretreatment, are formidable technological barriers which retard the development of a variety of industrial processes. Structural analyses of the corn fiber heteroxylans suggest that over 70% of the xylose backbone residues have one or more arabinose, 4-O-methyl glucuronic acid or other side chains (Figure 1). Therefore, it is hypothesized that commercial hemicellulase preparations could to be supplemented with accessory enzymes (enzyme cocktails) to increase effectiveness. Research is in progress to

Table VIII. Properties of Some Microbial Phenolic Acid Esterases

Organism	Molecular Weight	pI	Optimum Temperature (°C)	Optimum pH
Ferulic acid esterase				
Aspergillus oryzae (203)	30,000	3.6	45	4.5-6.0
Aspergillus niger (212) III	36,000	3.3	55-60	5
Streptomyces olivochromogenes (213)	29,000	7.9, 8.5	30	5.5
p-Coumaric acid esterase				
Neomallimastrix (214)	11,000	4.7	40	7.2

186

identify suitable xylanolytic enzyme preparations that can efficiently hydrolyze corn fiber xylan and xylooligosaccharides. Towards this end, Cotta has isolated an anaerobic bacterium (strain RZ) from rumen enrichment cultures which is capable of rapid growth on corn fiber xylan (Cotta, M. A., ARS-USDA, Peoria, IL, personal communication, 1998). Predictively, the strain produces a variety of xylanolytic enzymes, including endo-xylanase, β-xylosidase and α-arabinofuranosidase.

Literature Cited

1. Ward, O. P.; Moo-Young, M. *CRC Crit. Rev. Biotechnol.* **1989**, *8*, 237-274.
2. Wong, K. K. Y.; Tan, L.U. L.; Saddler, J. N. *Microbiol. Rev.* **1988**, *52*, 305-317.
3. Vikari, L.; Tenkanen, M.; Buchert, J.; Ratto, M.; Bailey, M.; Siika-aho, M.; Linko, M. In *Bioconversion of Forest and Agricultural Plant Residues*; Saddler, J. N., Ed.; CAB International: Oxford, **1993**; pp. 131-182.
4. Zeikus, J. G.; Lee, C.; Lee, Y.-E.; Saha, B. C. In *Enzymes in Biomass Conversion*; Leatham, G. F.; Himmel, M. E., Eds.; American Chemical Society: Washington, DC, **1991**; pp. 36-51.
5. Viikari, L.; Kantelinen, A.; Sundquist, J.; Linko, M. *FEMS Microbiol. Rev.* **1994**, *13*, 335-350.
6. Nissen, A. M.; Anker, L.; Munk, N.; Lange, N. K. In *Xylans and Xylanases*; Visser, J.; Beldman, G.; Kusters-Van Someren, M. A.; Voragen, A. G. J., Eds.; Elsevier Sciences Publishers: Amsterdam, **1992**; pp. 325-337.
7. Bothast, R. J.; Saha, B. C. *Adv. Appl. Microbiol.* **1997**, *44*, 261-286.
8. Hespell, R. B.; O'Bryan, P. J.; Moniruzzaman, M.; Bothast, R. J. *Appl. Biochem. Biotechnol.* **1997**, *62*, 87-97.
9. Moniruzzaman, M.; Dale, B. E.; Hespell, R. B.; Bothast, R. J. *Appl. Biochem. Biotechnol.* **1997**, *67*, 113-126.
10. Coughlan, M. P.; Hazlewood, G. P. *Biotechnol. Appl. Biochem.* **1993**, *17*, 259-289.
11. Saha, B. C.; Bothast, R. J. In *Fuels and Chemicals from Biomass*; Saha, B. C.; Woodward, J., Eds.; American Chemical Society: Washington, DC, **1997**; pp. 307-319.
12. McMillan, J. D. In *Enzymatic Conversion of Biomass for Fuel Production*; Himmel, M. E., Baker, J. O; Overend, R. P., Eds.; American Chemical Society: Washington, DC, **1993**; pp. 292-323.
13. Aspinall, G. O. In *The Biochemistry of Plants (A Comprehensive Treatise), Vol. 3. Carbohydrates: Structure and Function*, Preiss, J., Ed.; Academic Press: New York, NY, **1980**; pp. 473-500.
14. Kormelink, F. J. M.; Voragen, A. G. J. *Appl. Microbiol. Biotechnol.* **1993**, *38*, 688-695.
15. Shibuya, N.; Iwasaki, T. *Phytochemistry* **1985**, *24*, 285-289.
16. Gruppen, H; Hamer, R. J.; Voragen, A. G. J. *J. Cereal Sci.* **1992**, *16*, 53-67.
17. Doner, L. W.; Hicks, K. B. *Cereal Chem.* **1997**, *74*, 176-181.

18. Saulnier, L.; Marot, C.; Chanliaud, E.; Thibault, J.-F. *Carbohydr. Polym.* **1995**, *26*, 279-287.
19. Whistler, R.L.; Corbett, W. M. *J. Amer. Chem. Soc.* **1955**, *77*, 6328-6330.
20. Montgomery, R.; Smith, F.; Srivastava, H. C. *J. Amer. Chem. Soc.* **1956**, *78*, 2837-2839.
21. Montgomery, R.; Smith, F.; Srivastava, H. C. *J. Amer. Chem. Soc.* **1957**, *79*, 698-700.
22. Srivastava, H. C.; Smith, F. *J. Amer. Chem. Soc.* **1957**, *79*, 982-984.
23. Saulnier, L.; Vigouroux, J.; Thibault, J.-F. *Carbohydr. Res.* **1995**, *272*, 241-253.
24. Mueller-Hartley, I.; Hartley, R. D.; Harris, P. J.; Curzon, E. H. *Carbohydr. Res.* **1986**, *148*, 71-85.
25. Smith, M. M.; Hartley, R. D. *Carbohydr. Res.* **1983**, *118*, 65-80.
26. Timell, T. E. *Wood Sci. Technol.* **1967**, *1*, 45-70.
27. Wood, T. M.; Castanares, A.; Smith, D. C.; McCrae, S. I.; Brown, J. A. In *Xylans and Xylanases*; Visser, J.; Beldman, G.; Kusters-Van Someren, M. A.; Voragen, A. G. J., Eds.; Elsevier Sciences Publishers: Amsterdam, **1992**; pp. 187-202.
28. Poutanen, K.; Puls, J. In *Biogenesis and Biodegradation of Plant Cell Wall Polymers*, Lewis, G.; Paice, M., Eds.; American Chemical Society: Washington, DC, **1989**; pp. 630-640.
29. Bachmann, S. L.; McCarthy, A. J. *Appl. Environ. Microbiol.* **1991**, *57*, 2121-2130.
30. Lee, S. F.; Forsberg, C. W. *Can. J. Microbiol.* **1987**, *33*, 1011-1016.
31. Poutanen, K.; Tenkanen, M.; Korte, H.; Puls, J. In *Enzymes in Biomass Conversion*; Leatham, G. F. ; Himmel, M. E., Eds.; American Chemical Society: Washington, DC, **1991**; pp. 426-436.
32. Gilbert, H. J.: Hazlewood, G. P. *J. Gen. Microbiol.* **1993**, *139*, 187-194.
33. Filho, E. X. F.; Touhy, M. G.; Puls, J.; Coughlan, M. P. *Biochem. Soc. Trans.* **1991**, *19*, 25S.
34. Cotta, M. A. *Appl. Environ. Microbiol.* **1993**, *59*, 3557-3563.
35. Cotta, M. A. *Appl. Environ. Microbiol.* **1995**, *61*, 4396-4402.
36. Cotta, M. A.; Whitehead, T. R. *Curr. Microbiol.* **1998**, *36*, 183-189.
37. Leathers, T. D. *J. Ind. Microbiol.* **1989**, *4*, 341-348.
38. Leathers, T. D. *Appl. Environ. Microbiol.* **1986,** *52*, 1026-1030.
39. Berenger, J.-F.; Frixon, C.; Bigliardi, J.; Creuzet, N. *Can. J. Microbiol.* **1985**, *31*, 635-643.
40. Esteban, R.; Villanueva, R.; Villa, T. G. *Can. J. Microbiol.* **1982**, *28*, 733-739.
41. Yang, R. C. A.; MacKenzie, C. R.; Bilous, D.; Seligy, V. L.; Narang, S. A. *Appl. Environ. Microbiol.* **1988**, *54*, 1023-1029.
42. Comtat, J. *Carbohydr. Res.* **1983**, *118*, 215-231.
43. Dung, N. V.; Kamio, Y.; Abe, N.; Kaneko, J.; Izaki, K. *Appl. Environ. Microbiol.* **1991**, *57*, 445-449.

44.   Dung, N. V.; Vetayasuporn, S.; Kamio, Y.; Abe, N.; Kaneko, J.; Izaki, K. *Biosci. Biotech. Biochem.* **1993**, *57*, 1708-1712.
45.   Fujimoto, H.; Ooi, T.; Wang, S-L.; Takizawa, T.; Hidaka, H.; Murao, S.; Arai, M. *Biosci. Biotech. Biochem.* **1995**, *59*, 538-540.
46.   Simpson, H. D.; Haufler, U. R.; Daniel, R. M. *Biochem. J.* **1991**, *277*, 413-417.
47.   Winterhalter, C.; Liebl, W. *Appl. Environ. Microbiol.* **1995**, *61*, 1810-1815.
48.   Khasin, A.; Alchanati, I.; Shoham, Y. *Appl. Environ. Microbiol.* **1993**, *59*, 1725-1730.
49.   Sunna, A.; Prowe, S. G.; Stoffregen, T.; Antranikian, G. *FEMS Microbiol. Letts.* **1997**, *148*, 209-216.
50.   Kubata, B. K.; Suzuki, T.; Horitsu, H.; Kawai, K.; Takamizawa, K. *Appl. Environ. Microbiol.* **1994**, *60*, 531-535.
51.   Kubata, B. K.;Takamizawa, K., Kawai, K.; Suzuki, T.; Horitsu, H. *Appl. Environ. Microbiol.* **1995**, *61*, 1666-1668.
52.   Kormelink, F. J. M.; Searle-van Leeuwen, M. J. F.; Wood, T. M.; Voragen, A. G. J. *J. Biotechnol.* **1993**, *27*, 249-265.
53.   Li, X.-L.; Zhang, Z.-Q.; Dean, J. F. D.; Eriksson, K.-E.L.; Ljungdahl, L. G. *Appl. Environ. Microbiol.* **1993**, *59*, 3212-3218.
54.   Nakanishi, K.; Arai, H.; Yasui, T. *J. Ferment. Technol.* **1984**, *62*, 361-369.
55.   Alconada, T. M.; Martinez, M. J. *FEMS Microbiol. Lett.* **1994**, *118*, 305-310.
56.   Monti, R.; Terenzi, H. F.; Jorge, J. A. *Can. J. Microbiol.* **1991**, *37*, 675-681.
57.   Gomez de Segura, B.; Fevre, M. *Appl. Environ. Microbiol.* **1993**, *59*, 3654-3660.
58.   Mishra, C.; Keskar, S.; Rao, M. *Appl. Environ. Microbiol.* **1984**, *48*, 224-228.
59.   Tuohy, M. G.; Puls, J.; Claeyssens, M.; Vrsanskas, M.; Coughlan, M. P. *Biochem. J.* **1993**, *290*, 515-523.
60.   Tan, L.U. L.; Wong, K. K. Y.; Yu, E. K. C.; Saddler, J. N. *Enzyme Microb. Technol.* **1985**, *7*, 425-430.
61.   Tan, L.U. L.; Wong, K. K. Y.; Saddler, J. N. *Enzyme Microb. Technol.* **1985**, *7*, 431-436.
62.   Wong, K. K. Y.; Tan, L.U. L.; Saddler, J. N. *Can. J. Microbiol.* **1986**, *32*, 570-576.
63.   Wood, T. M.; McCrae, S. I. *Carbohydr. Res.* **1986**, *148*, 321-330.
64.   Haas, H.; Herfurth, E.; Stoffler, G.; Redl, B. *Biochim. Biophys. Acta* **1992**, *1117*, 279-286.
65.   Tenkanen, M.; Puls, J.; Poutanen, K. *Enzyme Microb. Technol.* **1992**, *14*, 566-574.
66.   Ujiie, M.; Roy, C.; Yaguchi, M. *Appl. Environ. Microbiol.* **1991**, *57*, 1860-1862.
67.   Panbangred, W.; Shinmyo, A.; Kinoshita, S.; Okada, H. *Agric. Biol. Chem.* **1983**, *47*, 957- 963.
68.   Barnier, R., Jr.; Desrocher, M.; Jurasek, L.; Paice, M. G. *Appl. Environ. Microbiol.* **1983**, *46*, 511-514.
69.   Lee, S. F.; Forsberg, C. W.; Rattray, J. B. *Appl. Environ. Microbiol.* **1987**, *53*, 644-650.

70. Matte, A.; Forsberg, C. W. *Appl. Environ. Microbiol.* **1992**, *58*, 157-168.

71. Magnuson, T. S.; Crawford, D. L. *Enzyme Microb. Technol.* **1997**, *21*, 160-164.

72. Tsujibo, H.; Miyamoto, K.; Kuda, T.; Minami, K.; Sakamoto, T.; Hasegawa, T.; Inamori, Y. *Appl. Environ. Microbiol.* **1992**, *58*, 371-375.

73. Stutzenberger, F. J.; Bodine, A. B. *J. Appl. Bacteriol.* **1992**, *72*, 504-511.

74. Sunna, A.; Puls, J.; Antranikian, G. *Biotechnol. Appl. Biochem.* **1996**, *24*, 177-185.

75. Thomson, J. A. *FEMS Microbiol. Rev.* **1993**, *104*, 65-92.

76. Sunna, A.; Antranikian, G. *Crit. Rev. Biotechnol.* **1997**, *17*, 39-67.

77. Withers, S. G. In *Carbohydrate Engineering*, Peterson, S. B.; Svensson, S.; Pederson, S., Eds.; Elsevier Science Publishers B. V.: Amsterdam, **1995**; pp. 97-113.

78. Tomme, P.; Warren, R. A. J.; Gilkes, N. R. *Adv. Microb. Physiol.* **1994**, *51*, 1-81.

79. Gilkes, N. R.; Henrissat, B., Kilburn, D. G.; Miller, R. C. *Microbiol. Rev.* **1991**, *55*, 303-315.

80. Henrissat, B. *Biochem. J.* **1991**, *280*, 309-316.

81. Gebler, J.; Gilkes, N. R.; Claeyssen, M.; Wilson, D. B.; Beguimn, P.; Wakarchut, W. W.; Kilburn, D. G.; Miller, R. C. J.; Warren, R. A. J.; Withers, S. G. *J. Biol. Chem.* **1992**, *267*, 12559-12561.

82. Sakka, K.; Kojima, Y.; Kondo, T.; Karita, S.; Ohmiya, K.; Shimada, K. *Biosci. Biotech. Biochem.* **1993**, *57*, 273-277.

83. Milward-Sadler, S. J.; Poole, D. M.; Henrissat, B.; Hazlewood, G. P.; Clarke, J. H.; Gilbert, H. *J. Mol. Microbiol.* **1994**, *11*, 375-382.

84. Christakopoulos, P.; Nerinckx, W.; Samyn, B.; Kekos, D.; Macris, B.; Van Beeumen, J.; Claeyssen, M. *Biotechnol. Letts.* **1996**, *18*, 349-354.

85. Sun, J. L.; Sakka, K.; Karita, S.; Kimura, T.; Ohmiya, K. *J. Ferment. Bioeng.* **1998**, *85*, 63-68.

86. Flint, H. J.; Martin, J.; McPherson, C. A.; Daniel, A. S.; Zhang, J.-X. *J. Bacteriol.* **1993**, *175*, 2943-2951.

87. Okada, H. *Adv. Protein Des.* **1989**, *12*, 81-86.

88. Campbell, R. L.; Rose, D. R.; Wakarchuk, W. W.; To, R.; Sung, W.; Yaguchi, M. In *Proceedings of the Second TRICEL Symposium on Trichoderma Cellulases and Other Hydrolases*; Suominen, P.; Reinikainen, T., Eds.; Foundation for Biotechnical and Industrial Fermentation Research: Helsinki, **1993**; pp. 63-72.

89. Torronen, A.; Rouvinen, J. *Biochemistry* **1995**, *34*, 847-856.

90. Kraulis, P. J. *J. Appl. Crystallogr.* **1991**, *24*, 946-950.

91. Godden, B.; Legon, T.; Helvenstein, P.; Pennickx, M. *J. Gen. Microbiol.* **1989**, *135*, 285-292.

92. Krengel, U.; Dijkstra, B. W. *J. Mol. Biol.* **1996**, *263*, 70-78.

93. Ko, E. P.; Akatsuka, H.; Moriyama, H.; Okada, H. *Biochem. J.* **1992**, 117-121.

94. Bray, M. R.; Clarke, A. J. *Eur. J. Biochem.* **1994**, *219*, 821-827.

95. Miao, S.; Ziser, L.; Aebersold, R.; Withers, S. G. *Biochemistry* **1994**, *33*, 7027-7032.
96. Wakarchuk, W. W.; Campbell, R. L.; Sung, W. L.; Davodi, J.; Yaguchi, M. *Protein Sci.* **1994**, *3*, 467-475.
97. Wong, K. K. Y.; Saddler, J. N. *Crit. Rev. Biotechnol.* **1992**, *12*, 413-435.
98. Whitehead, T. R.; Hespell, R. B. *Appl. Environ. Microbiol.* **1992**, *58*, 2764-2770.
99. Garcia-Campayo, V.; Wood, T. M. *Carbohydr. Res.* **1993**, *242*, 229-245.
100. John, M.; Schmidt, J. *Methods Enzymol.* **1988**, *160*, 662-671.
101. Yasui, T.; Matsuo, M., J. *Methods Enzymol.* **1988**, *160*, 696-700.
102. Lee, S. F.; Forsberg, C. W. *Appl. Environ. Microbiol.* **1987**, *53*, 651-654.
103. McCarty, A. J.; Ball, A. S.; Bachman, S. L. In *Biology of Actinomycetes*; Okami, Y.; Beppu, T. ; Ogawara, H., Eds.; Japan Scientific Societies Press: Tokyo, **1988**; pp. 283-287.
104. Buettner, R.; Bode, R. *J. Basic. Microbiol.* **1992**, *32*, 159-166.
105. Kumar, S.; Ramon, D. *FEMS Microbiol. Letts.* **1996**, *136*, 287-293.
106. Dobberstein, J.; Emeis, C. C. *Appl. Microbiol. Biotechnol.* **1991**, *35*, 210-215.
107. Nanmori, T.; Watanabe, T.; Shinke, R.; Kohno, A.; Kawamura, Y. *J. Bacteriol.* **1990**, *172*, 6669-6672.
108. Saxena, S.; Fierobe, H-P.; Gaudin, C.; Guerlesquin, F.; Belaich, J-P. *Appl. Environ. Microbiol.* **1995**, *61*, 3509-3512.
109. Almeida, E. M.de; Polizeli, M. de L. T. M.; Terenzi, H. F.; Jorge, J. A. *FEMS Microbiol. Letts.* **1995**, *130*, 171-176.
110. Zhu, H.; Cheng, K.-J.; Forsberg, C. W. *Can. J. Microbiol.* **1994**, *40*, 484-490.
111. Matsuo, M.; Fujie, A.; Win, M.; Yasui, T. *Agric. Biol. Chem.* **1987**, *51*, 2367-2379.
112. Manelius, A.; Dahlberg, L.; Holst, O. *Appl. Biochem. Biotechnol.* **1994**, *44*, 39-48.
113. Lachke, A. H.; Deshpande, M. V.; Srinivasan, M. C. *Enzyme Microb. Technol.* **1985**, *7*, 445-448.
114. Shao, W.; Weigel, J. *J. Bacteriol.* **1992**, *174*, 5848-5853.
115. Stutzenberger, F.; Bodine, A. B. *J. Ind. Microbiol. Biotechnol.* **1998**, *20*, 55-60.
116. Bachmann, S. L.; McCarthy, A. J. *J. Gen. Microbiol.* **1989**, *35*, 293-299.
117. Ximenes, F. A. de; Silveira, F. Q. P. de; Filho, X. F. *Curr. Microbiol.* **1996**, *33*, 71-77.
118. Poutenan, K.; Puls, J. *Appl. Microbiol. Biotechnol.* **1988**, *28*, 425-432.
119. Van Doorsler, E.; Kersters-Hilderson, H.; De Bruyne, C. K. *Carbohydr. Res.* **1985**, *140*, 342-346.
120. Matsuo, M.; Yasui, T. *Methods Enzymol.* **1988**, *160*, 684-695.
121. Herrmann, M. C.; Vrsanska, M.; Jurickova, M.; Hirsch, J.; Biely, P.; Kubicek, C. P. *Biochem. J.* **1997**, *321*, 375-381.
122. Rodionova, N. A.; Tavibilow, I. M.; Bezborodov, A. M. *J. Appl. Biochem.* **1983**, *5*, 300-312.
123. Oguntimein, G. B.; Reilly, P. J. *Biotechnol. Bioeng.* **1980**, 1143-1154.

124. Yasui, T.; Kizawa, H.; Masuda, Y.; Shinoyama, H. *Agric. Biol. Chem.* **1989**, *53*, 3381-3382.
125. Lachke, A. H. *Methods Enzymol.* **1988**, *160*, 679-684.
126. Sulistyo, J.; Kamiyama, Y.; Yasui, T. *Biosci. Biotech. Biochem.* **1994**, *58*, 1311-1313.
127. Sulistyo, J.; Kamiyama, Y.; Yasui, T. *J. Ferment. Bioeng.* **1995,** *79*, 17-22.
128. Henrissat, B.; Bairoch, A. *Biochem. J.* **1993**, *293*, 781-788.
129. Braun, C.; Meinke, A.; Ziser, L.; Withers, S. G. *Anal. Biochem.* **1994**, *212*, 259-262.
130. Armand, S.; Vielle, C.; Gey, C.; Heyraud, A.; Zeikus, J. G.; Henrissat, B. *Eur. J. Biochem.* **1996**, *236*, 706-713.
131. Greve, L. C.; Labavitch, J. M.; Hungate, R. E. *Appl. Environ. Microbiol.* **1984**, *47*, 1135-1140.
132. Poutanen, K. *J. Biotechnol.* **1988**, *7*, 271-282.
133. Ramon, D.; Veen, P. v. d.; Visser, J. *J. Biotechnol.* **1993**, *113*, 15-22.
134. Rombouts, F. M.; Voragen, A. G. J.; Searle-van Leeuwen, M. F.; Geraerds, C. C. J. M.; Schols, H. A.; Pilnik, W. *Carbohydr. Polym.* **1988**, *9*, 25-47.
135. Luonteri, E.; Siika-aho, M.; Tenkanen, M.; Vikari, L. *J. Biotechnol.* **1995**, *38*, 279-291.
136. Filho, E. X. F., Puls, J.; Coughlan, M. P. *Appl. Environ. Microbiol.* **1996**, *62*, 168-173.
137. Saha, B. C.; Bothast, R. J. *Appl. Environ. Microbiol.* **1998**, *64*, 216-220.
138. Komae, K.; Kaji, A.; Sato, M. *Agric. Biol. Chem.* **1982**, *46*, 1899-1905.
139. Hespell, R. B.; O'Bryan, P. J. *Appl. Environ. Microbiol.* **1992**, *58,* 1082-1088.
140. Kaneko, S.; Shimasaki, T., Kusakabe, I. *Biosci. Biotech. Biochem.* **1993**, *57*, 1161-1165.
141. Bezalel, L.; Shoham, Y.; Rosenberg, E. *Appl. Environ. Microbiol.* **1993**, *40*, 57-62.
142. Fernandez-Espinar, M. T.; Pena, J. L.; Pinaga, F.;Valles, S. *FEMS Microbiol. Letts.* **1994**, *115*, 107-112.
143. Kaji, A. *Adv. Carbohydr. Chem. Biochem.* **1984**, *42*, 383-394.
144. Uesaka, E.; Sato, M.; Raiju, M.; Kaji, A. *J. Bacteriol.* **1978**, *133*, 1073-1077.
145. Kaji, A., Sato, M.; Tsutsui, Y. *Agric. Biol. Chem.* **1981**, *45*, 925-931.
146. Tajana, E.; Fiechter, A.; Zimmermann, W. *Appl. Environ. Microbiol.* **1992**, *58*, 1447-1450.
147. Kaji, A.; Tagawa, K.*Biochim. Biophys. Acta* **1970**, *207*, 456-464.
148. Kormelink, F. J. M.; Searle-Van Leewan, M. J. F.; Wood, T. M.; Voragen, A. G. J. *Appl. Microbiol. Biotechnol* .**1991**, *35*, 753-758.
149. Van Laere, K. M. J.; Beldman, G.; Voragen, A. G. J. *Appl. Microbiol. Biotechnol.* **1997**, *47*, 231-235.
150. Manin, C.; Shareek, F.; Morosoli, R.; Kluepfel, D. *Biochem. J.* **1994**, *302*, 443-449.

151. Kaneko, S.; Sano, M.; Kusakabe, I. *Appl. Environ. Microbiol.* **1994**, *60*, 3425-3428.

152. Utt, E. A.; Eddy, C. K.; Keshav, K. F.; Ingram, L. O. *Appl. Environ. Microbiol.* **1991**, *57*, 1227-1234.

153. Whitehead, T. R.; Hespell, R. B. *J. Bacteriol.* **1990**, *172*, 2408-2412.

154. Weinstein, L.; Albershein, P. *Plant Physiol.* **1979**, *63*, 425-432.

155. Renner, M. J.; Breznek, J. A. *Appl. Environ. Microbiol.* **1998**, *64*, 43-52.

156. Puls, J.; Schmidt, O.; Granzow, C. *Enzyme Microb. Technol.* **1987**, *9*, 83-88.

157. Johnson, K. G.; Silva, M. C.; MacKemnzie, C. R.; Schneider, H.; Fontana, J. D. *Appl. Biochem. Biotechnol.* **1989**, *20/21*, 245-258.

158. Fontana, J. D.; Gebara, M.; Blumel, H.; Schneider, H.; MacKenzie, C. R.; Johnson, K. G. *Methods Enzymol.* **1988**, *160*, 560-571.

159. Ishihara, M.; Shimizu, K. *Mokazai Gakkaishi* **1988**, *34*, 58-64.

160. Dekker, R. F. H. *Biotechnol. Bioeng.* **1983**, *25*, 1127-1146.

161. Siika-aho, M.; Tenkanen, M.; Buchert, J.; Puls, J.; Vikari, L. *Enzyme Microb. Technol.* **1994**, *16*, 813-819.

162. Khandke, K. M.; Vithayathil, P. J.; Murthy, S. K. *Arch. Biochem. Biophys.* **1989**, *274*, 511-517.

163. Smith, D. C.; Forsberg, C. W. *Appl. Environ. Microbiol.* **1991**, *57*, 3552-3557.

164. Johnson, K. G.; Harrison, B. A.; Schneider, H.; MacKenzie, C. P.; Fontana, J. D. *Enzyme Microb. Technol.* **1988**, *10*, 403-409.

165. MacKenzie, C. R.; Bilous, D.; Schnieder, H.; Johnson, K. G. *Appl. Environ. Microbiol.* **1987**, *53*, 2835-2839.

166. Shao, W.; Obi, S. K. C.; Puls, J.; Weigel, J. *Appl. Environ. Microbiol.* **1995**, *61*, 1077-1081.

167. Puls, J; Schuseil, J. In *Hemicelluloses and Hemicellulases*; Coughlan, M. P.; Hazlewood, G. P., Eds.; Portland Press: London, **1993**; pp. 1-27.

168. Puls, J. In *Xylans and Xylanases*; Visser, J.; Beldman, G.; Kusters-Van Someren, M. A.; Voragen, A. G. J., Eds.; Elsevier Sciences Publishers: Amsterdam, **1992**; pp. 213-224.

169. Ishihara, M.; Inagaki, S.; Hayashi, N.; Shimizu, K. *Shinrin Sogo Kenkyusho Kenkyu Kokoku* **1990**, *359*, 141-157.

170. Biely, P.; Puls, J.; Schneider, H. *FEBS Lett.* **1985**, *186*, 80-84.

171. Khan, A. W.; Lamb, K. A.; Overend, R. P. *Enzyme Microb. Technol.* **1990**, *12*, 127-131.

172. Puls, J.; Poutenan, K. In *Enzyme systems for Lignocellulose Degradation*; Coughlan, M. P., Ed.; Elsevier: London, **1989**; pp. 151-165.

173. Poutenan, K.; Ratto, M.; Puls, J.; Vikari, L. *J. Biotechnol.* **1987**, *6*, 49-60.

174. Egana, L.: Gutierrez, R.; Caputo, V.; Peirano, A.; Steiner, J.; Eyzaguirre, J. *Biotechnol. Appl. Biochem.* **1996**, *24*, 33-39.

175. Lee, H.; To, R. J. B.; Latta, R. K.; Biely, P.; Schneider, H. *Appl. Environ. Microbiol.* **1987**, *53*, 2831-2834.

176. MacKenzie, C. R.; Bilous, D. *Appl. Environ. Microbiol.* **1988**, *54*, 1170-1173.

177. Biely, P.; MacKenzie, C. R.; Schneider, H. *Can. J. Microbiol.* **1988**, *32*, 767-772.

178. Poutenan, K.; Sundberg, M. *Appl. Microbiol. Biotechnol.* **1988**, *28*, 425-432.

179. Degrassi, G.; Okeke, B. C.; Bruschi, C. V.; Venturi, V. *Appl. Environ. Microbiol.* **1998**, *64*, 789-792.

180. Hazlewood, G. P.; Gilbert, H. J. In *Xylans and Xylanases*; Visser, J.; Beldman, G.; Kusters-Van Someren, M. A.; Voragen, A. G. J., Eds.; Elsevier Sciences Publishers: Amsterdam, **1992**; pp. 259-273.

181. Hespell, R. B.; O'Bryan-Shah, P. J. *Appl. Environ. Microbiol.* **1988**, *54*, 1917-1922.

182. Dupont, C.; Diagneault, N.; Shareck, F.; Morosoli, R.; Kluepfel, D. *Biochem. J.* **1996**, *319*, 881-886.

183. Lee, Y-E.; Lowe, S. E.; Zeikus, J. G. *Appl. Environ. Microbiol.* **1993**, *59*, 763-771.

184. Shao, W.; Weigel, J. H. *Appl. Environ. Microbiol.* **1995**, *61*, 729-733.

185. Luthi, E.; Jasmat, N. B.; Berguist, P. L. *Appl. Microbiol. Biotechnol.* **1990**, *34*, 214-219.

186. McCarthy, A. J.; Bachmann, S. L. In *Xylans and Xylanases*; Visser, J.; Beldman, G.; Kusters-Van Someren, M. A.; Voragen, A. G. J., Eds.; Elsevier Sciences Publishers: Amsterdam, **1992**; pp. 309-313.

187. McDermid, K. P.; Forsberg, C. W.; MacKenzie, C. R. H. *Appl. Environ. Microbiol.* **1990**, *56*, 3805-3810.

188. Margolles-Clark, E.; Tenkanen, M.; Soderlund, H.; Penttila, M. *Eur. J. Biochem.* **1996**, *237*, 553-560.

189. Gutierrez, R.; Cederland, E.; Hjelmqvist, L.; Peirano, A.; Herrera, F.; Ghosh, D.; Duax, W.; Jornvall, H.; Eyzaguirre, J. *FEBS Letts.* **1998**, *423*, 35-38.

190. Kormelink, F. J. M.; Lefebvre, B.; Stozyk, F.; Voragen, A. G. J. *J. Biotechnol.* **1993**, *27*, 267-282.

191. Halgasova, N.; Kutejova, E.; Timko, J. *Biochem. J.* **1994**, *298*, 751-755.

192. Sundberg, M.; Poutanen, K. *Biotechnol. Appl. Biochem.* **1991**, *13*, 1-11.

193. Bacon, J. S. D.; Chesson, A.; Gordon, A. H. *Agric. Environ.* **1981**, *6*, 115-126.

194. Grohmann, K. D.; Mitchell, J.; Himmel, M. E.; Dale, B. E.; Schroeder, H. A. *Appl. Biochem. Biotechnol.* **1989**, *20/21*, 45-61.

195. Biely, P.; MacKenzie, C. R.; Puls, J.; Schneider, H. *Bio/Technology* **1986**, *4*, 731-733.

196. Mitchell, D. J.; Grohmann, K.; Himmel, M. E. *J. Wood Chem. Technol.* **1990**, *10*, 111-121.

197. Biely, P.; Cote, G. L.; Kremnicky, L.; Weisleder, D.; Greene, R. V. *Biochim. Biophys. Acta* **1996**, *1298*, 209-222.

198. Biely, P.; Cote, G. L.; Kremnicky, L.; Greene, R. V.; Dupont, C.; Kluepfel, D. *FEBS Letts.* **1996**, *396*, 257-260.

199. Biely, P.; Cote, G. L.; Kremnicky, L.; Greene, R. V.; Tenkanen, M. *FEBS Letts.* **1997**, *420*, 121-124.

200. Poutanen, K.; Sundberg, M.; Korte, H.; Puls, J. *Appl. Microbiol. Biotechnol.* **1990**, *33*, 506-510.
201. Hartley, R. D.; Jones, E. C. *Phytochemistry* **1977**, *16*, 1531-1534.
202. Hartley, R. D.; Ford, C. W. In *Plant Cell wall Polymers: Biogenesis and Biodegradation*; Lewis, N. G.; Paice, M. G., Eds.; American Chemical Society: Washington, DC, **1989**; pp.137-145.
203. Tenkanen, M.; Schusewil, J.; Puls, J.; Poutanen, K. *J. Biotechnol.* **1991**, *18*, 69-83.
204. Balogun, S. B.; Gomes, J.; Steiner, W. *Food Technol. Biotechnol.* **1997**, *35*, 13-22.
205. McDermid, K. P.; MacKenzie, C. R.; Forsberg, C. W. *Appl. Environ. Microbiol.* **1990**, *56*, 127-132.
206. Borneman, W. S.; Hartley, R. D.; Morrison, W. H.; Akin, D. E.; Ljungdal, L. G. *Appl. Environ. Microbiol.* **1990**, *33*, 345-351.
207. Deobald, L. A.; Crawford, D. L. *Appl. Microbiol. Biotechnol.* **1987**, *26*, 158-163.
208. Donnelly, P. K.; Crawford, D. L. *Appl. Environ. Microbiol.* **1988**, *54*, 2237-2244.
209. Castanares, A.; McCrae, S. I.; Wood, T. M. *Enzyme Microb. Technol.* **1992**, *14*, 875-884.
210. Christov, L. P.; Prior, B. A. *Enzyme Microb. Technol.* **1993**, *15*, 460-475.
211. Bartolome, B.; Laulds, C. B.; Kroon, P. A.; Waldron, K.; Gilbert, H. J.; Hazlewood, G.; Williamson, G. *Appl. Environ. Microbiol.* **1997**, *63*, 208-212.
212. Faulds, C. B.; Williamson, G. *Microbiology* **1994**, *140*, 779-787.
213. Faulds, C. B.; Williamson, G. *J. Gen. Microbiol.* **1991**, *137*, 2339-2345.
214. Borneman, W. S.; Ljungdal, L. G.; Hartley, R. D.; Akin, D. E. *Appl. Environ. Microbiol.* **1991**, *57*, 2337-2344.

Chapter 12

Extruded Plastics Containing Starch and Chitin: Physical Properties and Evaluation of Biodegradability

K. A. Niño[1], S. H. Imam[2], S. H. Gordon[2], and L. J. G. Wong[1]

[1]Department of Microbiology and Immunology, University of Nuevo León, San Nicolás de los Garza, Nuevo León 64000, Mexico

[2]Biopolymer Research Unit, National Center for Agricultural Utilization Research, Agricultural Research Service, U.S. Department of Agriculture, 1815 N. University Street, Peoria, IL 61604

A blend containing 40% cornstarch, 25% low density polyethylene (LDPE), 25% ethylene-co-acrylic acid (EAA) and 10% urea as a plasticizer was prepared and extruded into blown plastic films. In another blend, half of the starch was replaced by chitin while the composition of LDPE, EAA and urea remained the same. Control films containing neat LDPE and EAA were also extruded. Mechanical properties and biodegradability of the films were investigated. These films were incubated with a consortium of bacteria (LD 76) in liquid culture to determine their biodegradability. Biodegradation was assessed by measuring film weight-loss, infrared absorption changes in FTIR spectra and by measuring changes in physical properties of the films, namely tensile strength and percent elongation, over a 30 day period. While starch/LDPE/EAA/urea films lost over 21% weight within 30 days, weight loss in starch-chitin/LDPE/EAA/urea was only 14.0% during the same time period. The controls, LDPE and EAA, showed negligible weight loss. FTIR spectra exhibited diminished bands corresponding to OH, C-O and Amide I region absorbances from the starch and chitin components. No spectral changes were observed in control samples. While percent elongation diminished significantly in both formulations, loss of tensile strength was less pronounced and somewhat variable.

The use of plastics in single-use, disposable applications has increased significantly, particularly in the second half of this century. The useful life of most plastics end after packaging is removed (1). Plastics persist in the environment due to their resistance to biodegradation. Improper disposal of plastics has been shown to cause harm to marine animals, including sea birds, turtles and other life forms (2). Recalcitrant plastics accumulate in the environment at a rate of 25 million tons per year (3). Plastic formulations containing starch from corn as a biodegradable component have been studied as part of an effort to increase the susceptibility of plastic products to biological degradative processes (4-6). Once disposed of in the environment, these materials are expected to biodegrade under the influence of environmental processes and eventually disintegrate into smaller and benign byproducts (7). Examples include agricultural mulch films, used for weed control and water retention in crops, as well as packaging materials and other rigid containers that have limited durability and can be discarded after use. Starch is a polysaccharide which has been used extensively in the development of biodegradable plastics. The objective of this study was to develop formulations containing other potentially useful natural polysaccharides, such as chitin, in starch-synthetic polymer blends and test their biodegradability under laboratory conditions.

Experimental

Plastics. The plastic films used in this study were obtained by the semi-dry method of Otey (8). Formulations are shown in Table I. Two test formulations were prepared, one containing 40% starch (dry-weight basis), while the other contained 20% starch and 20% chitin. Both formulations contained the same amount of LDPE, EAA and urea. Control films were neat LDPE or EAA, without any polysaccharide.

Table I. Plastic Formulations and Physical Properties

Formulation	LDPE (%)	EAA (%)	Starch (%)	Chitin (%)	Urea (%)	TS[a] (Mpa)	%E[b]
Starch/LDPE/ EAA/Urea	25	25	40	0	10	9.3	105.6
Starch/Chitin/ LDPE/EAA/ Urea	25	25	20	20	10	8.4	12.80
Neat LDPE	100					15.29	201.4
Neat EAA		100				17.73	234.8

[a] TS, tensile strength in Mega Pascal
[b] % E, percent elongation

Biodegradation Studies and Sample Analyses. Plastic films were cut into strips, 1.0 cm wide and 10.0 cm long. Strips were sterilized by soaking for 30 minutes in a 3% H_2O_2 solution, followed by several rinses with sterile distilled water, which also removed the urea. All strips were air-dried, numbered and initial weights were determined. Three strips were added to each culture flask containing 100 ml modified mineral medium and inoculated with 2 ml of an actively growing LD-76 consortium (4). Mineral medium contained KH_2PO_4, 0.4g; K_2HPO_4, 0.1g; $MgCl-7H_2O$, 0.2g; NaCl, 0.1g; CaCl, 0.02g; $FeCl_3$, 0.01g; NaMo, 0.002g; NH_4Cl, 1.5g; vitamin solution, 1.5 ml per liter of distilled water (4). The vitamin solution contained biotin, 0.2 mg; nicotinic acid, 2.0 mg; thiamine, 1.0 mg; p-aminobenzoic acid, 0.1 mg; pantothenic acid, 0.5mg; and pyridoxine, 5.0mg per 100 ml of distilled water. The pH was adjusted to 7.3 ± 0.1. Flasks were incubated at 30 °C with agitation (150 rpm) in a gyrotatory shaker for 45 days. Four flasks for each formulation were prepared. Samples representing each formulation were collected on day 7, day 15, day 30 and day 45. Degradation was assessed by the weight loss of strips, changes in the infrared spectra of the respective polymer using Fourier transform infrared (FTIR) spectroscopy and by measuring changes in two physical properties, tensile strength and percent elongation at break.

To determine the final percent weight loss, samples were carefully rinsed with distilled water to clean and exclude the material accumulated on the surface, and then air-dried at room temperature for 5 days before weighing.

For FTIR analysis, KBr pellets of air-dried strips were prepared following the method of Gould et al. (4) and spectra were obtained on a KVB/Analect RFX 75 FTIR Spectrometer. Mechanical properties of samples were determined with an Instron-Universal Testing Machine (Model 4201), using a gage length of 10 mm and a crosshead speed of 20 mm/min. Presented results are averages of three tests.

Results

Incorporation of chitin into the starch/LDPE/EAA/urea formulation had little effect on the tensile strength of the specimens. However, percent elongation was greatly impacted (Table I). Biodegradation studies revealed that initially, for the first fifteen days, the rate and extent of weight loss in both starch/LDPE/EAA/urea and starch-chitin/LDPE/EAA/urea films were quite similar (Figure 1). In both films, weight loss occurred at a rate of about 0.8%/day. Thereafter, films with starch continued to loose weight at a similar rate until day 30, when a plateau was reached. Whereas, films with chitin reached a plateau after only 15 days of incubation. The extent of weight loss in starch and starch-chitin films by day 45 was 23.1% and 14.0%, respectively (Figure 1). The weights of controls containing neat LDPE and EAA remained unchanged (Figure 1). This was expected, but illustrates the recalcitrant nature of conventional plastics.

FTIR analyses of the biodegraded films revealed large decreases in IR absorbance in the hydroxyl region (3400 cm^{-1}) and carbohydrate fingerprint region (960-1190 cm^{-1}) of the spectrum, which are absorbances attributed mostly to the polysaccharide component of these films (Figure 2). The formulation with added chitin showed decreased carbohydrate absorbances, as well as decreased absorbances in the Amide I region (1664 cm^{-1}) region of the spectrum, indicating that both polysaccharides were removed from the films due to microbial action. However, C-H stretching bands (2851-2921 cm^{-1}) and a C=O band (1705 cm^{-1}), indicitive of the synthetic polymers, remained unchanged for all formulations (Figure 2).

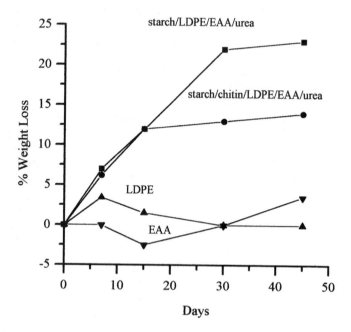

Figure 1. Weight loss in specimens as a result of exposure to cultures of LD-76 for 45 days.

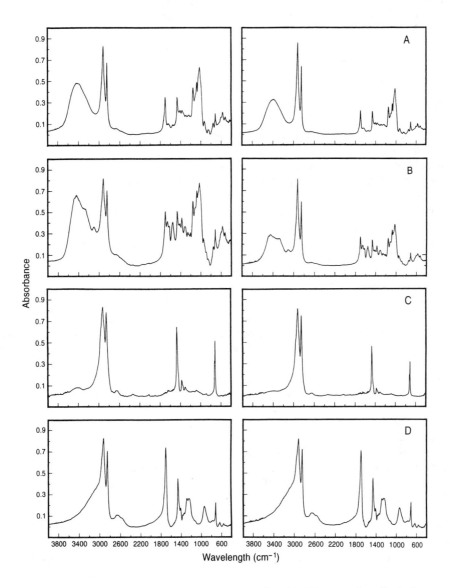

Figure 2. FTIR spectra of formulations exposed to LD-76 consortium for 0 day (left) and 30 days (right). (A) starch/LDPE/EAA (B) starch/chitin/LDPE/EAA, (C) Neat LDPE and (D) Neat EAA.

Generally, weight loss in films was accompanied by loss in percent elongation and tensile strength. Though tensile strength decreased in both formulations (starch/LDPE/EAA/urea and starch/chitin/LDPE/EAA/urea), the loss varied somewhat (3-23 %) between the two formulations (data not shown). On the other hand, decreases in percent elongation were more dramatic in these two samples (Figure 3). Some loss in percent elongation was also observed in control films, however, this loss was within the acceptable limits of varibility (Figure 3).

Discussion

Little is known about blending chitin with starch in LDPE/EAA blown films. These results indicate that blending chitin with starch reduces the tensile strength and impacts elasticity of the films, limiting this material to only a few practical applications. Importantly, both starch and chitin are accessible to microbial enzymes for breakdown, even though incorporation of chitin somewhat slowed the extent of biodegradation in these films. The weight loss in films due to biodegradation was generally accompanied by deterioration of mechanical properties. Starch films with no chitin showed substantial biodegradation as evidenced from the weight loss, FTIR and other physical data. Data also indicated that neither LDPE nor EAA underwent appreciable change. Gould and coworkers (4) found that 40% starch-containing LDPE/EAA blown films exposed in liquid culture lost close to 40% weight within 60 days of incubation. They too found that deterioration of mechanical properties accompanied the weight loss. Loss of starch in films, however never approach 100%. In this regard, investigators have also found that a portion of starch in both blown films (7) and injection molded samples (10) forms V-type complexes with EAA which are highly resistant to enzyme attack. Imam et al., (11) also found that in injection molded starch/PE composites, much of the starch is encapsulated within the PE thus making it inaccessible to hydrolytic enzymes.

The FTIR spectrometer is a useful tool to quantify the relative amount of both starch and chitin in synthetic polymer blends. In starch/LDPE/EAA films, weight losses obtained gravimetrically and by estimation from FTIR absorbance peaks appear to be quite similar. In fact, it was possible to measure precisely the percent loss of starch and chitin from the diminution of OH, C-O and Amide I absorbance peaks in FTIR spectra. Table II indicates a loss of 52.0 - 59.6% of the starch from starch/LDPE/EAA films. Whereas in starch-chitin/LDPE/EAA films, a modified absorbance ratio analysis (12) of FTIR spectra revealed losses of 48.9% of the starch and 14.1% of the chitin (Table II). Results indicate that while the synthetic polymers do not degrade in the plastic films studied here, it is clear that rapid and appreciable polysaccharide depletion leads to favorable deterioration of the mechanical properties of these films, leaving behind a weak plastic matrix susceptible to further physical disruption by a variety of biotic and abiotic factors. This work provides information on compounding and extrusion of more than one polysaccharide with synthetic polymers. Experiments are in progress to improve mechanical properties by using chitosan and other compatibilizing agents.

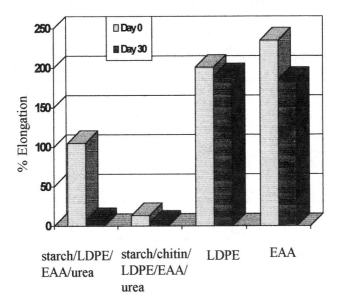

Figure 3. Change in mechanical properties (% elongation) of films after 30 days of exposure to LD-76 culture.

202

Acknowledgments

The authors thank Charles L. Swanson and Richard P. Westhoff for assistance in preparing extruded films and Christopher James for help in FTIR analyses.

Table II. Quantification of Starch and Chitin in 45 Day Biodegraded Samples from FTIR Absorbance and Weight Loss Data

Formulation	% Weight Loss	% Starch Loss	% Chitin Loss
Starch/ LDPE/EAA	23.10 [a] 26.50 [b]	51.98 [a] 59.63 [b]	—
Starch/Chitin/ LDPE/EAA	13.99 [c]	48.90 [c]	14.05 [c]

[a] Determined by gravimetric analysis.
[b] Calculated from OH (3400 cm^{-1}) absorbance ratios.
[c] Calculated from Amide I (1664 cm^{-1}) and C-O (1025 cm^{-1}) absorbance ratios.

Literature Cited

1. Ndon, U.J.; Levine, A.D.; Bradley B.S. *Sci. Tech.* **1992**, *26*, 2089.
2. Laist D. *Mar.Pollut. Bull.* **1987**, *18*, 305.
3. Byungtae, L.; Pometto III, A.L.; Fratzke, A.; Bailey Jr., T.B. *Appl. Environ. Microbiol.* **1991**, *57*, 678.
4. Gould, J.M., Gordon, S.H.; Dexter, L.B.; Swanson, C.L. *In Agricultural and Synthetic Polymers : Biodegradability and Utilization;* Glass, J.E.; Swift, G., Eds.; ACS Symposium Series No. 433; American Chemical Society: **1990**; pp. 65-75.
5. Imam, S.H.; Gould, J.M.; Gordon, S.H.; Kinney, M.P.; Ramsey, A.M.: Tosteson, T.R. *Curr. Microbiol.* **1992**, *25*, 1.
6. Arévalo-Niño,K; Sandoval, C.F.; Galan, L.J.; Imam, S.H.; Gordon, S.H.; Greene, R.V. *Biodegradation* **1996**, *7*, 231.
7. Greizerstein, H.B.; Syracuse, J.A.; Kosytyniak, P.J. *Polym. Degrad. Stabil.* **1993**, *39*, 251.

8. Otey, F.H.; Mark A.M. **1976**, U.S. Patent, 3,949,145.

9. Shogren, R.L.; Thompson A.R.; Greene, R.V.; Gordon S.H.; Cote, G.L. *J. Appl. Polym. Sci.* **1991**, *42*, 2279.

10. Imam, S.H.; Gordon, S.H.; Thompson, A.R.; Harry-O'kuru R.E.; Greene, R.V. *Biotechnol. Techn.* **1993**, *7*, 791.

11. Imam, S.H., Gordon, S.H.; Burgess-Cassler, A.; Greene, R.V. *J. Environ. Polym. Degrad.* **1995**, *3*, 107.

12. Gordon, S.H.; Imam, S.H.; Greene, R.V. *In The Polymeric Materials Encyclopedia: Synthesis, Properties and Applications;* Solamone J. Ed.; CRC Press: **1996**, Vol. 10(Q-S), pp. 7885-7892.

Chapter 13

Natural Systems for Better Bioremediation: Isolation and Characterization of a Phenanthrene-Utilizing Strain of *Alteromonas* from Guayanilla Coastal Water Southwest of Puerto Rico

Baqar R. Zaidi[1], Syed H. Imam[2], and Richard V. Greene[2]

[1]Department of Marine Sciences, P.O. Box 9013, University of Puerto Rico, Mayaguez, Puerto Rico 00681–9013
[2]Biopolymer Research Unit, National Center for Agricultural Utilization Research, Agricultural Research Service, U.S. Department of Agriculture, 1815 N. University Street, Peoria, IL 61604

Application of novel enzymes produced by bacteria, fungi or plants have tremendous potential for enhanced bioremediation strategies. Efforts to identify and manipulate enzymes with recombinant DNA methodology may generate microorganisms which efficiently degrade toxic chemicals. However, engineered catabolic traits will only lead to successful bioremediation if the host microorganism can be established in the environment. Those microorganisms that successfully overcome environmental stresses will greatly enhance our ability to remove pollutants. By studying mechanisms which allow naturally degradative isolates to become established and/or flourish in selected toxic environments, we hope to gain valuable insight and tools for future, more sophisticated bioremediation strategies. In this chapter, we describe isolation and identification of a bacterial strain capable of degrading phenanthrene. This bacterium was isolated from the Guayanilla Bay in Southwestern Puerto Rico. The strain belongs to the genus *Alteromonas* and showed 64% similarity to a previously characterized *Alteromonas communis* as determined by physiological and biochemical characterization. It was deposited in the NRRL Culture Collection and assigned accession number NRRL P2030. *Alteromonas* sp. NRRL P2030, as well as two other bacterial isolates from Puerto Rican coastal waters, degraded phenanthrene (10 μg/ml) 10 hr after addition to Guayanilla Bay seawater samples. Indigenous microbial flora in uninoculated samples lagged 40 hr before degrading phenanthrene. Addition of inorganic phosphate had no effect on the degradation of phenanthrene by NRRL P2030. Inorganic nitrogen, however, enhanced degradation over ten-fold, suggesting that the availability of nitrogen is important for phenanthrene degradation in natural waters.

Huge quantities of agricultural and industrial chemicals are produced and used in the industrialized world. Some of the chemicals are toxic, or become toxic after use in the industry, and are intentionally or unintentionally discharged into the environment. Additionally, chemicals produced by industrialized countries are shipped throughout the world, thus causing a global environmental problem. Microorganisms are principally responsible for the degradation of these chemicals. Some chemicals are readily degraded, but others require far longer durations. Slowly degrading chemicals accumulate in the environment and there is a need to identify methodology to increase their rates of degradation. One technique that holds considerable promise is environmental introduction of microorganisms to enhance biodegradation (bioremediation). Application of recombinant DNA methodology (biotechnology) to create microorganisms which more efficiently degrade toxic chemicals holds considerable promise. Several reports have documented enhanced degradation rates of toxic chemicals utilizing recombinant microbial strains. (*1-3*).

Biotechnology may create more efficient laboratory strains of microorganisms for bioremediation. However, recombinant strains will only be successful if they can cope with stresses associated with environments into which they are introduced. Isolation and identification of natural strains capable of degrading selected chemicals and subsequent evaluation of abiotic and biotic factors on degradation rates is of value for several reasons. First, in a bioremediation strategy, a well characterized isolate could be a useful tool by itself. Second, knowledge gained regarding factors affecting environmental performance of a natural isolate will likely be applicable to genetically engineered strains. Third, a naturally degradative isolate is an attractive option for improvement via recombinant means or as a resource of genetic material to be transferred to another host.

Polycyclic aromatic hydrocarbons (PAHs) are of great environmental concern because of their toxicity, potential for trophic magnification and resistance to biodegradation (*5*). Environments contaminated with large amounts of these fused-ring aromatic compounds are considered hazardous owing to potential carcinogenic, mutagenic and teratogenic effects of specific PAHs (*7*). Due to their mutagenic and carcinogenic properties, 16 PAHs have been listed as priority pollutants by the United States Environmental Protection Agency (USEPA). In the coastal marine environment, sources of PAH contamination include atmospheric deposition, the petrochemical industry, domestic and industrial wastewater, rivers, and spillage of petroleum or petroleum products from ships. Phenanthrene, although not particularly hazardous itself (*8*), is a good model compound for studying PAHs biodegradation because it is an intermediate product formed during the biodegradation of petroleum.

Over three million habitants occupy the small (110 by 35 miles) Caribbean Island of Puerto Rico. Coastal environments of Puerto Rico are prime repositories of PAHs because most of the industries as well as urban centers are located on the coast. For over 20 years Guayanilla, located in southwestern Puerto Rico, was the site of one of the biggest petrochemical complexes in the world. It is no longer in operation, having closed in 1982. However, the coastal environment of Guayanilla has not recovered from its 20-year exposure to hydrocarbon pollution. It is, therefore, an ideal study site to explore the apparent recalcitrant nature of these compounds in tropical coastal environments.

Our previous work with tropical fresh water samples from Puerto Rico, showed that indigenous microorganisms were more efficient than non-indigenous microorganisms in utilizing toxic compounds (*13a,b,c*). Therefore, a bacterial strain was isolated from Guayanilla Bay and evaluated for its ability to degrade phenanthrene. We describe here the physiological and biochemical characteristics of this isolate which indicate it is a species of *Alteromonas*. Previously described PAH-degrading bacteria mostly belong to the genus *Pseudomonas* (*3*), although these have typically isolated from soil (*12*). In this report, we investigate the kinetics of phenanthrene degradation mediated by the newly isolated *Alteromonas* sp., as well as the effect of selected environmental factors on the degradation kinetics.

Materials and Methods

Chemicals. Phenanthrene (>96% purity), agarose type VII (low gelling temperature) and cycloheximide were from Sigma Chemical Co., St. Louis, MO. Noble agar was from Difco Laboratories, Detroit, MI.

Environmental Samples. For isolation of bacteria capable of degrading phenanthrene, seawater samples were collected from Mayaguez, Guanica and Guayanilla coastal water. These samples were plated within 2 hr after collection. For biodegradation experiments, samples were collected from Guayanilla coastal water, stored with refrigeration and used within 2-3 weeks after collection.

Isolation of the bacterium. Seawater samples were filtered through 0.45 μm filters. Aliquots (0.1 and 0.2 ml) of the filtered samples were added to a 3.5 ml of seawater based Winogradsky mineral medium (*1*) containing 1% agarose at 30°C. To this medium, 5 ml of trace metal solution was immediately added along with 0.2 ml of an ethanolic solution of phenanthrene (8.5 mg/ml). Mixtures were vortexed and used to inoculate an underlayer as described below (*2*). The trace metal solution contained 200 ml of $FeSO_4 \cdot 7H_2O$, 10 mg of $ZnSO_4$, 3 mg of $MnCl_2 \cdot 4H_2O$, 20 mg of $CoCl_2 \cdot 6H_2O$, 1 mg of $CuCl_2 \cdot 2H_2O$, 2 mg of $NiCl_2 \cdot 6H_2O$, 500 mg of $Na_2MoO_4 \cdot 2H_2O$, and 30 mg of H_3BO_3 per liter of distilled water. After inocula contents were mixed, they were evenly distributed onto an underlayer of seawater based Winogradsky mineral medium that was solidified with 1.5% of Noble agar and contained 0.1 g of cycloheximide (dissolved in minimal amount of acetone) per liter to discourage fungal growth. Plates were incubated at 30°C. Phenanthrene degrading colonies were identified by a halo of clearing in the opaque phenanthrene layer. Isolates were repeatedly grown in marine broth and transferred back to medium containing phenanthrene as the sole source of carbon to ensure that degradative capability was retained. The bacterium was deposited in the NRRL Culture Collection (Peoria, IL) as accession number NRRL P2030.

Phenanthrene Biodegradation

Respirometer. Biodegradation was determined by measuring CO_2 production in a closed circuit Micro-Oxymax respirometer (Columbus Instrument International Corporation, Columbus, OH). All tests and controls were run in duplicate in 250-ml respirometer sample chambers. Sample chambers were placed in a water bath at a constant temperature of 25°C. Tests were conducted with 20-ml sterile or non-sterile seawater containing phenanthrene (10 μg/ml) and the selected isolate as indicated in the text. Non-sterile seawater with phenanthrene but without added bacteria or non-sterile seawater without added bacteria and without phenanthrene were used as controls. Sample chambers were automatically refreshed and the total accumulation of CO_2 resulting from biodegradation was recorded at every 6 hr.

UV Spectrophotometer. In parallel to respirometer studies, experiments were conducted in which phenanthrene degradation was also determined by the diminution of its absorption peak (250 nm) in a Hitachi-200 spectrophotometer. *Alteromonas* sp. NRRL P2030 (1 x 10^5 cells/ml) was added to samples of sterile seawater mineral medium containing 1 μg/ml of phenanthrene. The bacterium was not added to controls.

Inorganic Nutrients. To determine the effects of inorganic N and P on the degradation of phenanthrene by *Alteromonas* sp. NRRL P2030, duplicate samples of sterile seawater were amended with 10 μg/ml of phenanthrene and 100 μg/ml of KNO_3 or K_2HPO_4, and were inoculated as described in the text.

Results

Isolation and Identification of Bacterial Strain. An indigenous bacterial strain was isolated from coastal waters of Puerto Rico and was characterized by Gram stain, cellular morphology, colony morphology and metabolic and biochemical identification tests (Table I). This bacterium stained Gram-negative. In a 20-24 hr culture, bacteria were motile. Thereafter, they shed flagella and became non-motile. The bacterium was also oxidase positive. Cultures grew slowly at 15°C, grew well between 20-40°C, and 30°C was the optimum temperature for growth. Morphologically, cells were curved rods about 2 μm long. The bacterium was obligatory marine, requiring seawater for growth. When streaked on Marine agar, colonies were not distinctive, having smooth edges and a creamy color. Cultures did not survive well at 4°C, precluding storage at that temperature.

No single organic compound serves as a principal source of carbon and energy for species of *Alteromonas*. They are able to utilize from 20-58 organic compounds as sole or principal sources of carbon, including pentoses, hexoses, disaccharides, sugar acids, sugar alcohols, amino acids and aromatic compounds (*12*). Eleven diagnostic traits (boldface type in Table I) allowed for the preliminary identification of the unknown isolate and were the following: cell shape, pigmentation, negative growth at 4°C,

Table I. Differential characteristics of the species of the genus *Alteromonas*[a]

	A. espejiana	*A. communis*	*Alteromonas* sp. NRRL P2030
Cell shape	St	Cu	Cu
Luminescence	-	-	-
Oxidase	+	+	+
Growth at:			
4°C	-	-	-
35°C	d	+	+
40°C	-	+	+
Utilization of:			
D-Mannose	d	+	+/+
D-Galactose	+	d	+/+
D-Fructose	d	+	-/-
Sucrose	+	-	d/+
Maltose	+	d	+/+
N-Acetylglucosamine	-	-	+/-
D-Mannitol	+	+	+/+
Glycerol	-	+	+/+
Pigmentation	-	-	-/-
L-Arabinose	-	d	d/-
D-Glucose	+	+	+/+
L-Rhamnose	-	d	d/d
D-Xylose	-	d	d
Citrate	+	+	+/-
Galacturonate	-	d	d/-
Acetate	+	+	+/-
Propionate	+	d	+/d

Table I. Continued.

	A. espejiana	A. communis	Alteromonas sp. NRRL P2030[b]
Utilization of:			
Aconitate	+	+	+
m-Inositol	-	+	-/-
Adonitol	-	-	-/-
L-Alanine	+	+	+/+
D-Alanine	- .	+	-/-
L-Serine	+	+	+/d
D-Serine			d/-
L-Leucine	d	-	+/d
L-Aspartate	+	d	+/d
L-Glutamate	-	+	+/d
L-Proline	d	+	d/-
L-Phenylalanine	d	d	d/-
L-Threonine	+	-	+/d
L-Ornithine			d/-
L-Pyroglutamic acid			d/-
D,L-Carnitine			d/-
γ-Aminobutyric acid			d/-
Inosine			+/d
Uridine			+/-
Thymidine			d/-
Putrescine			d/-
2-Aminoethanol			d/-
2,3-Butanediol			d/-
D,L-α-Glycerol phosphate			d/-
Glucose-6-phosphate			d/-
Glucose-1-phosphate			d/-

Continued on next page.

Table I. Continued.

	A. espejiana	A. communis	Alteromonas sp. NRRL P2030[b]
Utilization of:			
Bromosuccinic acid			+/-
Succinamic acid			+/-
Glucuronamide			+/-
Alaninamide			+/+
L-Alanylglycine			+/+
L-Asparagine			+/d
Glycogen			+/+
Tween 40			+/+
Tween 80			+/+
L-Aspartic acid			+
L-Glutamic acid			+
α-Ketoglutaric acid			d
D,L-Lactic acid			d
Malonic acid			d
Propionic acid			+
Quinic acid			d
D-Saccharic acid			d
Cis-Aconitic acid			+
D-Galactonic acid lactone			d
D-Galacturonic acid			d
D-Gluconic acid			d
D-Glucosaminic acid			d
D-Glucuronic acid			d

Table I. Continued.

	A. espejiana	A. communis	Alteromonas sp. NRRL P2030[b]
Utilization of:			
β-Hydroxybutyric acid			+
γ-Hydroxybutyric acid			+
D-Melibiose			+
β-Methyl-D-glucoside			+
D-Raffinose			d
Psicose			d
D-Trehalose			+
Turanose			+
Methylpyruvate			d
D-Galactose			d
Gentiobiose			+
α-D-Glucose			+
α-Lactose			+
α-D-Lactose lactulose			+
α-Cyclodextrin			d
Dextrin			+
Antibiotic Sensitivity			
Streptomycin			S
Spectinomycin			S
Kasugamycin			R

[a]Boldface type indicates traits useful for the preliminary identification of species: St, straight rods; and Cu, curved rods.

[b]Not able to utilize the following organic compounds as sole source of carbon and energy: L-rhamnose, D-sorbitol, xylitol, i-erythritol, mono-methylsuccinate, D-fructose, L-fucose, N-acetyl-D-galactosamine, D-arabitol, cellobiose, formic acid, α-hydroxybutyric acid, uronic acid, phenylethylamine, glycyl-L-aspartic acid, p-hydroxyphenylacetic acid, itaconic acid, α-ketobutyric acid, α-ketovaleric acid, sabacic acid, phenol, *p*-nitrophenol.

positive growth at 35°C, and utilization of D-galactose, D-maltose, D-mannitol, D-mannose, sucrose, glycerol and N-acetylglucosamine. These eleven trains indicated that the isolate was either *A. espejiana* or *A. communis*. All 113 traits were compared with the eleven known species of *Alteromonas*, the isolate showed highest similarity (64%) to *A. communis*, and lowest similarity (25%) to *A. aurantia*. The similarity to *A. espejiana* was 62% and, therefore, the isolate was not classified at the species level.

Phenanthrene Degradation. When microorganisms aerobically degrade or mineralize organic compounds, carbon dioxide is produced. There is a positive correlation between the amount of CO_2 released and the extent of degradation of a carbon substrate (*13*). Thus, the level of CO_2 measured by a respirometer is a good indicator of both the extent and rate of degradation. When phenanthrene was added (10 μg/ml) to the seawater samples from Guayanilla Bay and inoculated with *Alteromonas* sp. NRRL P2030, phenanthrene degradation was significantly enhanced (Figure 1). After an acclimation period of less than 10 hr, the degradation rate was quite high and most of the phenanthrene was degraded within 30 hr. In the absence of *Alteromonas* sp. NRRL P2030, indigenous microbes present in seawater samples required more than 50 hr to degrade the same concentration of phenanthrene, primarily due to an extended lag period of roughly 40 hr. Seawater samples without phenanthrene were also examined as a negative control. Expectedly, some CO_2 production was observed, presumably due to catabolism of endogenous carbon sources by the indigenous microbial flora.

The degradation of phenanthrene by *Alteromonas* sp. NRRL P2030 was specifically confirmed spectrophotometrically. Phenanthrene-containing seawater exhibited a major absorbance peak at 250 nm (A_{250}) and lesser ones at 292 nm and 273 nm. Figure 2 shows that A_{250} is stable in sterile seawater, while about a 70% diminution of this absorption is observed in 96 hr if the sample is inoculated with 1×10^5 cells/ml of *Alteromonas* sp. NRRL P2030. Also, at 96 hr cell counts rose by two orders of magnitude (not shown), indicating that *Alteromonas* sp. NRRL P2030 could utilize phenanthrene as a growth substrate. Controls lacking phenanthrene exhibited only slightly increased cell numbers.

In other systems, PAH degradation is achieved much more rapidly in mineral media as opposed to the natural environment (*14*). One possible explanation for this may be lack of essential nutrients. Particularly in seawater, inorganic N and P are known to be limiting factors. Therefore, experiments were conducted in which Guayanilla Bay water containing phenanthrene and *Alteromonas* sp. NRRL P2030 was supplemented with KNO_3 or K_2HPO_4. Samples with added P accumulated CO_2, indicative of phenanthrene degradation, at a rate that was similar to controls with no additions (Figure 3). However, in seawater samples with added N, phenanthrene degradation (CO_2 accumulation) was increased by several orders of magnitude. Samples lacking phenanthrene accumulated little CO_2 even when N was added, validating the assay. These results strongly suggest that the availability of N in Guayanilla Bay waters is crucial for achieving efficient degradation of PAHs and other related substrates.

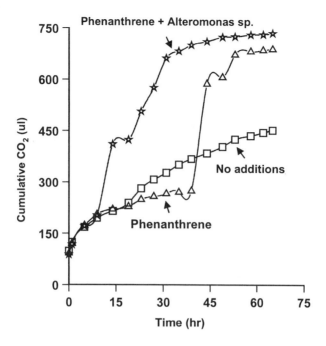

Figure 1. Biodegradation of phenanthrene (10µg/ml) in Guayanilla water.

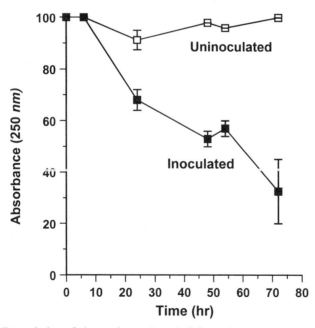

Figure 2. Degradation of phenanthrene (1 µg/ml) in sterile seawater medium by *Alteromonas* sp. NRRL P2030.

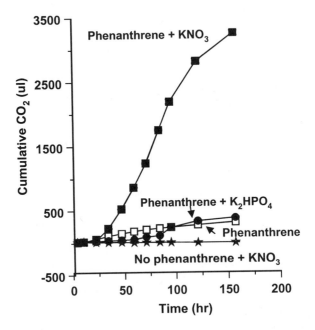

Figure 3. Effect of nitrate and phosphate on degradation of phenanthrene (10 μg/ml) by *Alteromonas* sp. NRRL P2030 in sterile Guayanilla Bay water.

Discussion

In this study, a bacterial strain capable of utilizing phenanthrene as a sole source of carbon and energy was isolated from Guayanilla, Puerto Rico, coastal water and was identified as either *Alteromonas communis* or *Alteromonas espejiana*. It was deposited in the NRRL Culture Collection under accession number P2030. Several methods were evaluated for isolation of phenanthrene degraders. For example, spraying of an ethanolic solution of phenanthrene (*15*) on seawater based mineral medium plates both before and after incubation. However, the overlay technique described by Bogardt et al. (*16*) was found to be more effective because a clearing zone around a particular colony established that it was phenanthrene degrader. This methodology allowed for the isolation from various Puerto Rican coastal environments of several, yet to be characterized, bacterial strains that were capable of relatively rapid phenanthrene degradation (Figure 4). Initially, it took about 3 weeks at room temperature for colonies to grow in seawater based mineral medium supplemented with phenanthrene, but subsequently bacteria grew within 3-4 days. These strains were later grown on Zobell marine agar and retested positively for the ability to degrade phenanthrene. Isolates, however, were found to be particularly sensitive to temperature below 5 °C.

Alteromonas sp. NRRL P2030 also catabolizes Tween-80 (Table I), a nonionic surfactant. Surfactants are often used to disperse oil following major petroleum spills in marine water (*14*). The utilization of both Tween-80 and phenanthrene by *Alteromonas* sp. NRRL P2030, has important implications. By simultaneously catabolizing PAHs and Tween-80 the bacterium would remove surfactant as well as pollutant from the environment. In this regard, further investigations are planned with other surfactants that should provide practical information pertaining to the use of *Alteromonas* sp. NRRL P2030.

Degradation of PAHs occurs mainly from microbial action and the environment may greatly impact this process. In this regard, Shiari (*17*) has studied PAH biodegradation in oxidized surficial sediments and concluded that factors such as salinity, season, temperature and ambient PAH concentration influence the process. Availability of inorganic nutrients is yet another important factor which may influence PAH degradation. Zaidi et al. (*18*) found that the concentration of inorganic P, but not N, in lake water was a limiting factor for the biodegradation of low concentrations of *p*-nitrophenol. Conversely, in this study it appears the concentration of inorganic N, but not P, in Puerto Rico coastal seawater limits the biodegradation of phenanthrene (Figure 3). Utilizing *Alteromonas* sp. NRRL P2030, studies are in progress to further examine the effects of nutrient concentration and other biotic factors on phenanthrene degradation in seawater.

Acknowledgments

We thank Jan V. Lawton, Biopolymer Research Unit, National Center for Agricultural Utilization Research, Peoria, IL, for her assistance in biodegradation assays and

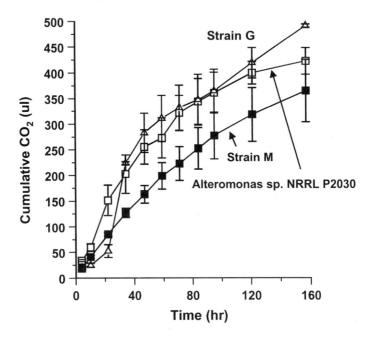

Figure 4. Degradation of phenanthrene (10 μg/ml) in sterile seawater mineral medium by bacteria isolated from Guayanilla Bay (*Alteromonas* sp. NRRL P2030), Mayaguez Bay (strain M) and Guanica coastal water (strain G).

Helen Gasdorf, Microbial Properties Research Unit, National Center for Agricultural Utilization Research, Peoria, IL, for her assistance in bacterial identification. This research was supported in part by the Office of the Associate Dean of Research, Faculty of Arts and Sciences, University of Puerto Rico, Mayaguez, Puerto Rico.

Literature Cited

1. Adams, R.H.; Huang, C.M.; Higson, F.K.; Brenner, V.; Focht, D.D. *Appl. Environ. Microbiol.* **1992**, *58*, 647-654.
2. Bruhn, C.; Bayly, R.C.; Knackmuss, H.J. *Arch. Microbiol.* **1988**, *150*, 171-177.
3. Latorre, J.; Reineke, W.; Knackmuss, H.J. *Arch. Microbiol.* **1984**, *140*, 159-165.
4. Maccubbin, A.E.; Black, P.; Trzeciak, L.; Black, J.J. *Bull. Environ. Contm. Toxicol.* **1985**, *34*, 876-882.
5. Muller, J.G.; Chapman, P.J.; Blattmann, B.O.; Pritchard, P.H. *Appl. Environ. Microbiol.* **1990**, *56*, 1079-1086.
6. Phillips, D.H. *Nature* (London) **1983**, *303*, 468-472.
7. Zaidi, B.R.; Imam, S.H. *J. Gen. Appl. Microbiol.* **1996**, *42*, 249-256.
8. Zaidi, B.R.; Metah, N.; Imam, S.H.; Greene, R.V. *Biotechnol. Lett.* **1996**, *18*, 565-570.
9. Zaidi, B.R.; Imam, S.H.; Greene, R.V. *Current Microbiol.* **1996**, *33*, 292-296.
10. Chakrabarty, A.M. *Annu. Rev. Genet.* **1976**, *10*, 7-30.
11. Yen, K.M.; Gunsalus, I.C. *Proc. Natl. Acad. Sci. USA* **1982**, *79*, 874-878.
12. Baumann, P.; Gautheir, M.J.; Baumann, L. In *Bergy's Manual of Systematic Bacteriology*; Kreig, N.R.; Holt, J.G. (Eds.); **1972**, Vol. 1; pp. 343-352.
13. Zaidi, B.R.; Stucki, G.; Alexander, M. *Environ. Toxicol. Chem.* **1988**, *7*, 143-151.
14. Manilal, V.B.; Alexander, M. *Appl. Microbiol. Biotechnol.* **1991**, *35*, 401-405.
15. Guerin, W.F. *Microb. Ecol.* **1989**, *17*, 89-104.
16. Bogardt, A.H.; Hemmingsen, B.B. *Appl. Environ. Microbiol.* **1992**, *58*, 2579-2582.
17. Shiari, M.P. *Appl. Environ. Microbiol.* **1989**, *55*, 1391-1399.
18. Zaidi, B.R.; Murakami, Y.; Alexander, M. *Environ. Sci. Technol.* **1988**, *22*, 1419-1425.

Chapter 14

Microbial Degradation of a Poly(lactic acid) as a Model of Synthetic Polymer Degradation Mechanisms in Outdoor Conditions

A. Torres[1], S. M. Li[2], S. Roussos[3], and M. Vert[2]

[1]GIRSA Research Center, 52000 Lerma, Mexico
[2]URA-CNRS 1465, CRBA, Faculty of Pharmacy, 34060 Montpellier, France
[3]Laboratory of Biotechnology, ORSTOM, 34032 Montpellier, France

Poly(α-hydroxy alkanoates) are known to easily degrade hydrolitically in aqueous media. In the case of lactic acid polymers (PLA), microbial degradation mechanisms are not well known. The fate of a PLA in the presence of microorganisms was investigated. First, the assimilation of by-products, such as monomer, lactic acid, dimer and oligomers by two microorganisms, a fungus (*Fusarium moniliforme*) and a bacterium (*Pseudomonas putida*) was evaluated in aqueous culture media. Second, 1 cm^2 plates of a racemic PLA (Mw=40,000) were allowed to age in an aqueous medium containing a mixed culture of the same microorganisms. Finally, racemic PLA plates were buried in a wood and recovered after 8 weeks to incubate them in laboratory conditions for 8 weeks more. Results showed that by-products of PLA chemical degradation can be assimilated by microorganisms thus demonstrating that PLA can be considered as a bioassimilable polymer. In contrast, it was observed that PLA plates must be initially degraded by chemical hydrolysis until the formation of such by-products and then the microorganisms are able to assimilate them, suggesting that PLA can not be considered as a biodegradable polymer.

Polymer materials developed during the last 50 years are generally resistant to microbial attack, a property that is now regarded as a problem with respect to solid waste management. As a source of alternative environment friendly material, the synthesis of degradable polymers is becoming of great interest. Among the few polymer families which have been identified so far as good candidates, aliphatic polyesters are the most attractive materials. These polymers, also named poly(hydroxy alkanoates), can be divided into two subgroups: poly(β-hydroxy alkanoates), such as poly(β-hydroxy butyrate) (PHB) or higher analogues and their copolymers; and poly(α-

218

hydroxy alkanoates), such as poly(lactic acid) (PLA), poly(glycolic acid)(PGA) and their copolymers.

Poly(β-hydroxy alkanoates) are microbial polymers and their biodegradation has been well documented (*1,2*). In contrast, the biodegradability of PLA/GA polymers is still questioned, although these polymers do resorb from compost and *in vivo* after chemical degradation. It has been observed indeed that chemical hydrolysis of PLA/GA polymers yields oligomers that can be released to the aqueous medium once they become small enough to solubilize. The final hydrolysis products are then lactic and glycolic acids (*3-7*). Whether or not these by-products can be totally assimilated by microorganisms has not been well demonstrated.

Results dealing with the fate of PLA polymers under controlled or natural conditions are presented. First, two microorganisms able to use lactic acid or PLA by-products as their only source of carbon and energy were selected: a filamentous fungus, *Fusarium moniliforme*, and a bacterium, *Pseudomonas putida*. Second, the ability of selected microorganisms to assimilate PLA by-products, such as lactic acid (LA), lactyl lactic acid (dimer) and higher oligomers, was tested to demonstrate the PLA bioassimilation. The effect of enantiomeric composition on the assimilation rate was also studied by using two series of compounds: racemic and L-pure lactic acid derived forms. Finally, to analyze possible PLA biodegradation, i.e., direct microbial attack on high molecular weight polymer chains, racemic PLA (PLA50) plates were aged under controlled conditions in a mixed culture and in the natural soil.

Screening of Microorganisms

A screening of filamentous fungi strains was realized by using three different substrates: D,L-lactic acid, racemic PLA oligomers (Mw=1000) and a PLA/GA copolymer (Mw=150,000). The first two substrates were used in a mineral liquid culture, whereas the copolymer was used on a solid culture (mineral agar medium). Racemic oligomers and the copolymer were added to the media after sterilization to avoid chemical degradation. Selection criteria in liquid culture were: total lactic acid consumption, residual nitrogen concentration and final biomass production, after 7 days at 28°C in shaken flasks (150 rpm).

Only three of 14 fungi strains tested were able to use D,L-lactic acid as their only source of carbon and energy: two strains of *F. moniliforme* and one strain of *Penicillium roqueforti*. Besides, *F. moniliforme* strains totally assimilated the nitrogen source [$(NH_4)_2SO_4$] and produced the highest biomass concentration (3 g/l of dry weight mass). These three strains could also assimilate racemic oligomers.

In the case of the PLA/GA copolymer, only one *F. moniliforme* strain was able to grow at the surface after two months of incubation at 28°C.

It was then concluded that *F. moniliforme* was able to assimilate PLA. However, it was not yet determined whether this assimilation occurred because of an enzymatic activity or because of a chemical hydrolysis followed by consumption of by-products.

Furthermore, a strain of *Pseudomonas putida* was also selected as a microorganism able to consume these kinds of compounds, on the basis of its capability to assimilate a large variety of products.

Bioassimilation of PLA By-Products

The ability of selected microorganisms to assimilate racemic and L-pure lactic acid compounds was tested in liquid cultures containing lactic acid, dimer and oligomers. In all cases, pure cultures of each microorganism were used and only in the case of oligomers was a mixed culture of both tested.

Dimers and oligomers were added aseptically after media sterilization. D,L- and L oligomers were 2000 and 4000 Mw, respectively. Lactic acid and dimer consumption were evaluated by direct HPLC of the liquid media. Figure 1 shows a typical HPLC chromatogram. It can be seen that the method detected not only lactic acid (17.205 min retention time), but also dimer and trimer (15.722 and 14.988 min retention time, respectively).

In the case of oligomers, media were heterogeneous because some of them were not soluble. The soluble fraction was then evaluated by alkaline hydrolysis of a small homogeneous volume, followed by HPLC quantification, and the total residual oligomers were measured as lactic acid after complete alkaline hydrolysis of the whole flask volume at different culture times (Figure 2).

Results showed that D,L and L monomers were totally assimilated by the two microorganisms without any influence of enantiomeric composition. Differences appeared with dimer assimilation: *F. moniliforme* started to consume dimers earlier than *P. putida* and the L-form was better assimilated (Figure 3).

In contrast, D,L-oligomers were consumed faster than the L-form. It was interesting to note that the mixed cultures of both microorganisms had a synergistic effect on the D,L-oligomer assimilation: After 8 days of incubation, 80% of oligomers were consumed, whereas only 30% of them were consumed by the bacteria and 20% by the fungi in the same period (Figure 4).

It was then supposed that both microorganisms possessed a complementary enzymatic system which permitted a faster assimilation of D,L-oligomers in a mixed culture. *P. putida* would be able to produce enzymes that hydrolyze chains randomly (endo-enzymes) and *F. moniliforme* would only be able to produce enzymes that attack the chain ends (exo-enzymes).

In the case of L-oligomers, solubility was very poor in all media and general assimilation was limited, suggesting that chain stereoregularity restricted the microbial activity.

In order to compare chemical and microbial degradation, a control without microorganisms was analyzed for dimer- and oligomer-containing media.

Chemical hydrolysis was then observed by the gradual accumulation of typical by-products: the soluble fraction of oligomers, which includes monomer (lactic acid), dimer, trimer, tetramer, pentamer, etc. Besides, it was noted that these soluble

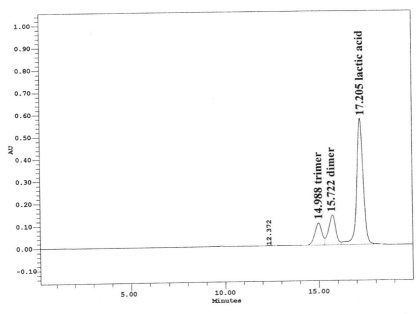

Figure 1. Typical HPLC analysis showing monomer (lactic acid), dimer and trimer.

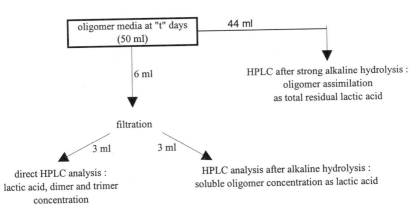

Figure 2. Protocol used to analyze oligomer heterogeneous media.

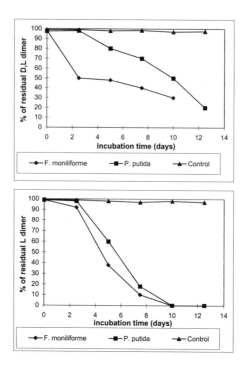

Figure 3. Assimilation of D,L- and L-dimers by the two microorganisms.

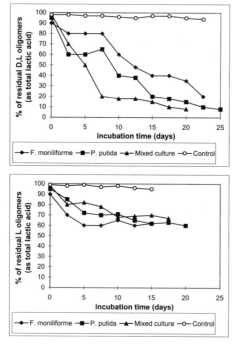

Figure 4. Assimilation of D,L- and L-oligomers by the two microorganisms in pure cultures and in a mixed cultures.

oligomers were released into the control media slower than in the microbial media only in the case of D,L-oligomers (Figure 5).

A synergistic effect can be observed again in mixed cultures containing the D,L-oligomers. The soluble fraction increases and then decreases faster than in the pure cultures, thus showing that there exists an acceleration of oligomer hydrolysis and then a faster assimilation. It can be suggested then that the assimilation rate for racemic oligomers depends on the enzymatic activity and not on the chemical hydrolysis rate.

In contrast, the soluble fraction of L-oligomers is very small for all cases. A gradual accumulation in control media is observed, but in a very low rate. In the microbial media, an even lower concentration of soluble oligomers is observed, suggesting that, for L-oligomers, the assimilation rate would be limited indeed by the chemical degradation rate.

At this point, it can be concluded that the PLA hydrolysis by-products, especially the racemic ones, can be totally bioassimilated provided that suitable microbial populations are present. However, the bioassimilation of L-oligomers would be retarded and the enzymatic activity would be inhibited by crystallinity.

Degradation Test of PLA

Chemical degradation of racemic PLA (PLA50) in the bulk state is an autocatalytic phenomenon that produces a differentiation between the inner and the outer part of bulky specimens, with the inner part being degraded faster than the surface (8). The effect of this chemical degradation phenomenon on mechanical properties of PLA specimens has been well documented (9-10).

Parallelepiped plates (10X10X2 mm) of a PLA50 (Mw=40,000) were aged under different conditions to analyze a possible biodegradation and to compare it with the chemical degradation.

In a first experiment, PLA50 plates were incubated at 30°C in shaking flasks (150 rpm) containing aqueous culture media incubated with *F. moniliforme* and *P. putida*. Until 17 weeks, water absorption rate of plates were the same in microbial and in control cultures. After this period, plates of microbial media absorbed water very much faster: 800% in weight at 32 weeks vs. control media. Typical differential degradation was also observed at 17 weeks, but after that the difference between center and surface disappeared and at 32 weeks samples had completely burst. In control media, differentiation remained during the whole experimentation time.

In a second experiment, PLA50 plates were buried in soil and recovered after 8 weeks. A second incubation period of 8 weeks was realized at 30°C in Petri dishes containing mineral agar medium. At the end of 16 weeks total degradation time, specimens showed that surfaces were invaded with mycelia growth (Figure 6).

Some filaments were even able to penetrate into the bulky mass, suggesting that microorganisms could reach the oligomers eventually accumulated during the first degradation period.

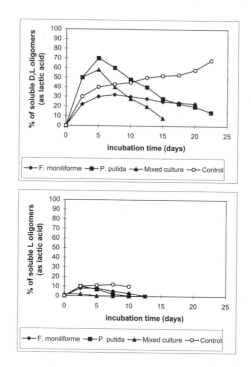

Figure 5. Soluble fraction of oligomers in the different media.

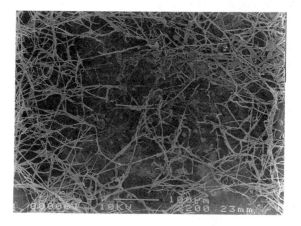

Figure 6. SEM micrograph showing mycellia growth on racemic PLA plates.

Five fungal strains were isolated from specimens and their ability to assimilate racemic oligomers was evaluated in mixed cultures in the same way as *F. moniliforme* and *P. putida*. Soil strains appeared to be more efficient than these two microorganisms in assimilating racemic oligomers. PLA bioassimilation under natural conditions was therefore confirmed (*11*).

It was inferred that racemic PLA plates were initially degraded by chemical hydrolysis and then, after some degradation extent, microorganisms were able to assimilate the by-products formed. The way in which the microorganisms assimilated these compounds was a function of water availability, as suggested by the different degradation mechanisms observed for the plates incubated in liquid culture media and for those buried in the soil and then incubated on solid agar media where water is limited.

Conclusion

According to our findings, lactic acid polymers in the bulk state may be considered as bioassimilable but not as biodegradable materials. This means that PLA would be degraded by some microorganisms but only after some degree of chemical hydrolysis, once oligomers are formed. These conclusions are of great interest because we can now establish the difference between bioassimilation and biodegradation phenomena.

The time frame for total degradation of PLA (bioassimilation + chemical hydrolysis) will depend on several factors: molecular weight, type of specimens (films, powder, plates, etc.), enantiomeric composition (related to crystallinity), microbial capacity, environment conditions (humidity, temperature, pH), etc.

Literature Cited

1. Doi, Y.; Kumagai, Y.; Tanashi, N.; Mukai, K. In *Biodegradable Polymers and Plastics;* Vert, M.; Feijen, J.; Albertsson, A.; Scott, G.; Chiellini, E., Eds.; Royal Soc. Chem.: London, **1992**; pp. 139-146.
2. Jendrossek, D.; Knoke, T.; Habibian, R. B.; Steinbuchel, A.; Shlegel, H. G. *J. Environ. Polymer Deg.* **1993**, *1(1),* 53-63.
3. Fukuzaki, H.; Yoshida, M.; Asano, M.; Kumakura, M. *Eur. Polym. J.* **1989**, *25(10),* 1019-1026.
4. Li, S. M.; Garreau, H.; Vert. M. *J. Mat. Med.* **1990**, *1,* 123-130.
5. Zhu, J. H.; Shen, Z. R.; Wu, L. T.; Yang, S. L. *J. Appl. Polym. Sci.* **1991**, *43,* 2099-2106.
6. Therin, M.; Christel, P.; Li, S. M.; Garreau, H.; Vert, M. *Biomaterials* **1992**, *13(9),* 594-600.
7. Grizzi, I.; Garreau, H.; Li, S. M.; Vert, M. *Biomaterials* **1995**, *16,* 305-311.
8. Li, S. M.; Vert, M. In *Degradable Polymers*; Scott, G.; Gilead, D. Eds.; Chapman & Hall: London, **1995**; pp. 43-87.

9. Li, S. M.; Garreau, H.; Vert, M. *J. Mat. Sci.: Materials in Medicine* **1990**, *1*, 198.

10. Li, S. M.; Garreau, H.; Vert, M. *J. Mat. Sci.: Materials in Medicine* **1990**, *1*, 131.

11. Torres, A.; Li, S. M.; Roussos, S.; Vert, M. *J. Appl. Polym. Sci.* **1996**, *62*, 2295-2302.

APPLICATIONS

Chapter 15

Membranes of Cellulose Derivatives as Supports for Immobilization of Enzymes

R. Lagoa[1], D. Murtinho[2], and M. H. Gil[1]

[1]Department of Chemical Engineering, University of Coimbra, 3000 Coimbra, Portugal
[2]Department of Chemistry, University of Coimbra, 3000 Coimbra, Portugal

Membranes were prepared from cellulosic derivatives (cellulose acetate, cellulose propionate and cellulose acetate butyrate). Their chemical and physical properties were determined by DSC, water vapor sorption and contact angle evaluation. Catalase, alcohol oxidase and glucose oxidase were covalently linked to these membranes and catalytic activity and stability were examined. The activity results of the immobilization and the stability of the coupled enzymes were found to correlate well with the studied properties of the supports. Cellulose acetate membranes yielded the most active conjugate of support and enzyme. Highly hydrophobic membranes from propionate and butyrate esters of cellulose yielded lower activities, but better storage stability.

Immobilized enzymes, proteins and cells offer advantages because they can be easily handled and recovered; i.e., easily removed from reaction mixtures and repeatedly used in continuous processes. In this way, expensive enzymes can be economically recycled for industrial and clinical applications. To be of value, enzymes attached to a support must retain some of their initial activity. The activity retained by the immobilized enzyme is dependent on a large number of parameters, including the coupling method, the enzyme used and the nature of the support.

Due to their good chemical and physical properties, cellulosic materials have been widely used as supports for enzyme immobilization (1,2). Examples include membranes for hemodialysis (3), for reverse osmosis (4), for microfiltration (5), in biosensors (1) and as chromatographic supports (6).

In our group we have been interested in suitable supports for the fixation of various biological compounds. We have used polyethylene (7), cellulose and pectin (8), and agar (9) based graft copolymers, as well as cellulosic materials for the immobilization of enzymes, proteins and cells. Simultaneously, we have studied the possibility of applying these systems to industry and in medicine.

In this paper, we report the immobilization of enzymes of importance in biosensor technology onto well characterized membranes derived from cellulose derivatives.

Experimental

Materials. Cellulose propionate (CP) and acetate-butyrate (CAB) were purchased from Aldrich (Dorset, UK) and cellulose acetate (CA) from Sigma (Dorset, UK). Glucose oxidase (β-D-glucose:oxygen 1-oxireductase; EC 1.1.3.4) from *Aspergillus niger*, catalase (H_2O_2:H_2O_2 oxireductase; EC 1.11.1.6) from bovine liver and alcohol oxidase (alcohol:oxygen oxireductase; EC 1.1.3.13) from *Candida boidinii* were supplied by Sigma. Sodium periodate, hexamethylene diamine and glutaraldehyde were obtained from Merck (Darmstadt, Germany). All other reagents were of analytical grade.

Preparation of the Membranes. Solutions of cellulose derivatives in tetrahydrofuran (10% w/v) were cast on a glass plate (20x20 cm), with the aid of a casting knife (0.33 mm). The solvent was evaporated at room temperature.

Determination of Water Vapor Sorption. Prepared membranes were conditioned over a saturated solution of copper sulfate (98% R.H.) at 25°C, until constant weight was achieved. Afterwards, the polymer was, in each instance, weighed every minute for ten minutes. The initial sorption capacity was obtained graphically after extrapolation to zero time. Each sample was dried to constant mass, under reduced pressure, at 100°C. The percentage of water uptake is given by:

% sorption = $(M_i - M_f)/ M_f \times 100$

where M_i is the initial mass at zero time and M_f is the final dry mass.

Determination of Static Contact Angles. Static contact angles were determined with a Contact-θ-Meter (Livereel, Durham, England). A sessile drop of water was placed on the surface of the membranes and the contact angle was determined at 20°C.

Characterization of the Membranes by DSC. Differential scanning calorimetric analyses were carried out on the membranes using a Polymer Laboratories, PL-DSC analyzer. Samples of membranes, weighing 5-6 mg, were sealed in aluminum pans. The samples were heated at a rate of 10°C min^{-1} in a N_2 gas purged atmosphere, using a flow rate of 10 cm^3 min^{-1}.

Enzyme Immobilization. Membrane sections were allowed to react with 10 cm^3 sodium periodate (0.5 M) for 2 hr in the dark. They were then treated with 10 cm^3 of a 1% hexamethylene diamine (w/w) solution for 18 hr. This step was followed by immersion for 2 hr at 4 °C in 10 cm^3 of 5% glutaraldehyde (v/v) in 0.05 M phosphate buffer at pH 7.5. Membranes were thoroughly washed with distilled water between each step (*10*). Enzyme solutions (glucose oxidase, 9 cm³, 2 mg/cm³; alcohol oxidase, 15 cm³, 4 mg/cm³; and catalase 15 cm³, 2 mg/cm³, all in 0.05 M phosphate buffer, pH 7.5) were allowed to couple to activated membranes over a period of 20 hr at 4°C. The enzyme-coupled membranes were thoroughly washed with 0.05 M phosphate buffer, pH 7.5, and stored in the same buffer at 4°C until activity assays were performed.

Enzyme Activity Assays. The activity of immobilized glucose oxidase (GOX) and free glucose oxidase was determined by the *o*-dianisidine procedure (*11*) (1 IU = 1 mmol D-glucose oxidized per minute). Free enzyme solution (0.05 cm³) was added to a buffered reaction mixture (50 mM acetate, pH 5.1) that consisted of 2.4 cm³ of 0.21 mM *o*-dianisidine, 0.5 cm³ of 10% D-glucose and 0.1 cm³ of 60 U/cm³ peroxidase. In the case of the immobilized enzyme, because the chromogen adsorbs onto the membranes, the process was separated into two steps. A weighed membrane portion containing immobilized glucose oxidase was immersed in 5 cm³ D-glucose (10% solution in 50 mM acetate buffer, pH 5.1). Then, 0.1 cm³ aliquots were withdrawn from the reaction vessel and reacted with 2.4 cm³ of 0.21 mM *o*-dianisidine and 0.1 cm³ 60 U/cm³ peroxidase.

The activity of free alcohol oxidase (AOD) and of immobilized alcohol oxidase was determined by the 4-aminoantipirine procedure (*12*), using 50 mM phosphate buffer and methanol as the substrate (1 IU = 1 mmol methanol oxidized per minute). Free alcohol oxidase (0.1 cm³) was added to 1 cm³ of 6 U/cm³ peroxidase and 1.9 cm³ of an indicator solution that consisted of 0.4 mM 4-aminoantipirine, 11 mM phenol and 1.5 M methanol. In the case of immobilized enzyme, weighed portions of membranes were immersed in 3 cm³ of peroxidase solution and 6 cm³ of indicator solution.

The activity of immobilized catalase was determined by following the decomposition of H_2O_2 at 240 nm (1 IU = 1 mmol of H_2O_2 degraded per minute). A weighed portion of membrane was immersed in 15 cm³ of 100 mM H_2O_2 in 50 mM phosphate buffer, pH 7.0.

Storage Stability of Free and Immobilized Glucose Oxidase. The storage stability of glucose oxidase was assessed. A 0.2 mg/cm³ of free enzyme solution, in 50 mM acetate buffer, pH 5.1, as well as membranes containing immobilized enzyme in the same buffer, were stored at 4°C for 3 days. Residual activity was then determined.

Results and Discussion

Membranes from cellulose acetate, cellulose propionate and cellulose butyrate were characterized by determination of water vapor sorption, thermal properties and static contact angle. The obtained results are presented in Table I. Among the tested materials, cellulose acetate has the highest water sorption capacity and the lowest contact angle. This can be explained by an increase in hydrophobic character due to the alkyl groups of propionate. Cellulose acetate butyrate presents two considerations:

a) the increased alkyl chain length of butyrate groups and b) the presence of two substituent groups.

Some of the thermal characteristics of the various cellulosic ester membranes are also given in Table I. All the membranes have a Tg. Cellulose acetate has the highest Tg, probably because this polymer is more ordered than cellulose propionate or cellulose acetate butyrate. Cellulose propionate and cellulose acetate butyrate are amorphous membranes.

Table I - Physical Characteristics of the Cellulosic Membranes

Membrane	Water Sorption Capacity (%)	Contact Angle (°)	T_g (°C)	ΔH (cal/g)	Tm (°C)
Cellulose Acetate	13.6±0.5	73±3	191.2	0.49	210.6
Cellulose Propionate	7.7±0.1	78±3	152.2	-	-
Cellulose Acetate butyrate	4.6±0.4	79±3	173.8	6.23	-

These three membranes were used to covalently link catalase, glucose oxidase and alcohol oxidase. The enzymes were covalently linked to the supports by use of sodium periodate as reported by Gil *et al.* (*10*). Relative activities are shown in Figure 1. The data in the graph suggest that the behavior of the enzyme-linked-membrane is not dependent on the enzyme, but rather on the support and that cellulose acetate membranes better couple the tested enzymes than do cellulose propionate or cellulose acetate butyrate.

These results can be related to the physical characteristics indicated in Table I. More hydrophilic membranes allow a better yield of coupled activity due to a better interaction between the enzyme solution and the support. On the other hand, the access of the enzyme to cellulose propionate and cellulose acetate butyrate membranes may be more difficult due to steric problems, as well as due to the higher substitution of the hydroxyl groups. Apparently, enzyme activities are strongly dependent upon the characteristics of the support.

The stability of GOX was investigated by monitoring the activity after 3 days of storage at 4°C (Table II). In comparison with the initial activities, supports which yield greater activity are those that offer less stabilization to the enzyme. Thus, enzymes immobilized onto CA membranes are less stable than those linked to CP and CAB membranes. This implies that a hydrophobic microenvironment reduces enzyme hydration and also contact with the aqueous solution, thus diminishing conformational mobility which leads to denaturation.

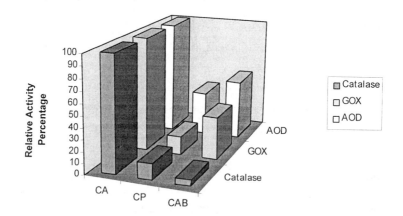

Figure 1. Relative activity of various enzymes covalently linked to various cellulosic supports.

Table II. Storage Stability of Free (IU / g Protein) and Immobilized (IU / g Membrane) Glucose Oxidase

Membrane	Day 0 Activity (IU/g)	Day 3 Activity (IU/g)	Activity Retention (%)
Free enzyme	0.144	21.4×10^{-3}	14
Cellulose acetate	1.9	266×10^{-3}	14
Cellulose propionate	0.3	236×10^{-3}	79
Cellulose acetate butyrate	0.7	239×10^{-3}	34

The stability of CP-immobilized GOX is distinctly greater as a result of specific interactions between the enzyme and support. Alternatively, processes which could lead to denaturation may have become diffusion limited. Immobilized biocatalytic systems, in which activity is limited by diffusion of the reactants, can have an apparent stability which is greater than the stability of the free enzyme molecules.

Conclusions

Characteristics, such as crystalinity, hydrophilicity and permeability, can be correlated with the activity and stability of enzymes immobilized onto membranes. Membranes of cellulose derivatives offer a good model for the study of the factors which influence the behavior of immobilized enzyme systems. Data from this model indicate that fusion enthalpies and water sorption values are useful in establishing practical parameters for the creation of effective supports for enzyme immobilization. Contact angle phenomena are of limited importance.

Acknowledgments

The authors gratefully acknowledge the financial support provided by JNICT (Portugal) for R. Lagoa.

Literature Cited

1. Gil, M. H.; Alegret, S.; Alves da Silva, M.; Alegria, A. C.; Piedade, A. P. In *Cellulosics: Materials for Selective Separations and Other Technologies*; Kennedy, J. F.; Philips, G. O.; William, P. A., Eds.; Ellis Horwood Ltd.: New York, NY, **1993**; pp. 163-171.
2. *Enzymes in Industry and Medicine*; Bickerstaff, G. F. Ed.; Whitstable Litho: Whistable, Kent, **1987**; 1st Ed.

3. Ohno, M.; Suzuki, M; Miyagi, M.; Yagi, T.; Sakurai, H.; Ukai, T. In *Cellulosics: Chemical, Biochemical and Material Aspects*; Kennedy, J. F.; Philips, G. O.; William, P. A., Eds.; Ellis Horwood Ltd.: New York, NY, **1993**; pp. 415-420.

4. Koros, W. J.; Fleming, G. K.; Jordon, S. M.; Kim T. H.; Hoehn, H. H. *Prog. Polym. Sci.* **1988**, *13*, 339-340.

5. Manabe, S.; Kamata, Y.; Iijima, H.; Kamide, K., *Polym. J.* **1987**, *19*, 391-404.

6. *Enzyme Technology*, Chaplin, M. F.; Bucke, C., Eds.; Cambridge University Press: **1990**; 1st Ed., Chapter 2.

7. Ramos, M. C.; Gil, M. H.; Garcia, A. P.; Cabral, J. M. S.; Guthrie, J. T. *Biocatalysis* **1992**, *6*, 223-234.

8. Beddows, C. G.; Gil, M. H.; Guthrie, J. T. *Polymer Bulletin* **1984**, *11*, 1-6.

9. Alves da Silva, M.; Beddows, C. G.; Gil, M. H.; Guthrie, J. T.; Guiomar, A. J.; Rotov, S.; Piedade, A. P. *Radiat. Phys. Chem.* **1990**, *35*, 98-101.

10. Gil, M. H.; Piedade, A. P.; Alegret, S.; Orellana, A. *Bios. and Biol.* **1992**, *7*, 645-652.

11. Huggett, A. G.; Nixon, D. A. *Lancet* **1957**, *2*, 368.

12. *Biosensors - A Practical Approach*; Cass, A. E. G., Ed.; IRL Press-Oxford University Press: Oxford, **1990**; pp. 117.

Chapter 16

Reducing Soil Erosion Losses with Small Applications of Biopolymers

William J. Orts and Gregory M. Glenn

Western Regional Research Center, Agricultural Research Service, U.S. Department of Agriculture, 800 Buchanan Street, Albany, CA 94710

High molecular weight, synthetic polyacrylamides (PAM) are used increasingly in the field to prevent erosion because, as explored in this study, they are relatively large, water soluble polymers which flocculate readily with soil due to charge affinity or Van der Waals attraction. A lab-scale erosion test was established to screen biopolymer solutions for a similar efficacy in reducing shear-induced erosion. In lab-scale furrows, chitosan, starch xanthate, and acid-hydrolyzed cellulose microfibrils proved viable in reducing suspended solids in the run-off water from test soil. For all of the polymers tested, erosion of clay-rich soils is reduced by increasing the concentration of exchangeable calcium. Optimization of biopolymer properties to improve their economic competitiveness in this application is discussed.

Soil erosion is a major problem threatening agricultural production and contaminating surface waters with overruns of pesticides and fertilizers. The magnitude of soil erosion is especially extensive in arid areas where soils lack the structure to withstand the shearing action of running water. For example, in the arid parts of the Pacific Northwest (Washington, Oregon, and Idaho), approximately 1.5 million hectares are surface irrigated, with erosion losses of 5 to 50 tons of soil per ha per year (1). The soil in this region is often derived from lava and ash and lacks polysaccharide stabilizing materials. Run-off water from these fields contributes to the build-up of silt in rivers, such as the Snake River in Idaho.

One effective tool for reducing erosion during irrigation is the addition of small quantities of polyacrylamide (PAM) to the in-flowing water (1-5). Lentz *et al.* (1) added 5-20 ppm of (anionic) PAM in the first 1-2 hours of the furrow irrigation of highly erodible soil and reduced sediment losses by up to 97% compared to untreated soil. This represented an ideological breakthrough in the use of soil conditioners. By adding the conditioner to the water, only the soil at the surface is treated, improving the cohesiveness in the ~1mm thick layer at the soil surface. Thus, only small quantities of polymer are required compared to the traditional practice of adding soil conditioners to "all" of the soil in the cultivation layer. Previously, hundreds and even thousands of pounds of soil conditioner per hectare were required to effectively control erosion losses.

Although the benefits of PAM are clearly seen by farmers, there are some considerations associated with this synthetic polymer which merit further review. One concern is that the long-term environmental impact of PAM is not known. PAM is a synthetic polymer that was not designed to biodegrade. It is very stable in soil-exposure environments, with "slow" deamination reported as its primary route to biodegradation (6, 7). Although PAM has been used in waste treatment facilities for many years, little is known about its long-term environmental effect when used in an agricultural environment. A second consideration is that acrylamide, the monomer used to synthesize PAM, is toxic, and any appreciable monomer impurity in the product polymer must be avoided. This problem has been suitably addressed by suppliers who provide PAM with only minute traces of monomer (8). It is imperative that producers continue to maintain a high standard of PAM in an economic environment of rapid expansion. Already, more than 700,000 acres are treated with PAM per year (9), with that number increasing by hundreds of thousands of acres per year. Third, at its relatively high price of $3.50 - $4.50/lb, treatment for one year costs between $15 and $40 per acre. Any reduction in this price would be a clear advantage to farmers. Finally, it is not clear that PAM works equally effectively, or even at all, on certain soil types. Wallace and Wallace (5) showed that soils with specific charge/pH profiles require a different charge balance in solution than that provided by commercial PAM alone. Having an "arsenal" of polymers with an array of soil-stabilizing mechanisms would provide farmers greater flexibility in controlling erosion.

In this report, we will discuss some of the functional attributes of PAM that make it an effective erosion control agent. Then, we will use this information to explore other polymers that may also be utilized in this role. Since PAM is a synthetic polymer that was not designed to achieve both biodegradability and functional performance, we will focus on biodegradable biopolymers, with long-term environmental impacts that may be better understood. There is a wide array of potentially suitable biopolymers with the primary functional attributes of PAM. They disperse readily in water, have a high molecular weight, and carry a charge. In this report, we use data from a lab-scale soil furrow test to screen several biopolymers for their efficacy in controlling soil erosion.

Materials and Methods

Polyacrylamides. Polymers from a variety of sources were screened for their efficacy in controlling erosion. Polyacrylamide, PAM, samples were either linear homopolymers or linear acrylic acid copolymers, with varying molecular weights, solubilities, and charges. Samples kindly provided by Cytec Industries of Stamford,

CT, were trademarked Magnifloc with product codes 835A, 836A, 837A, 846A, 905N, 442C, 492C, and 494C, and also included several relatively short-chain samples specially provided for this project (*10*). Samples kindly provided by Allied Colloids, Suffolk, VA, were trademarked Percol with product codes 338, E24, and LT25. Several PAM samples ranging in molecular weight from 200,000 to 15 million were purchased from Aldrich. All molecular weight and charge data were determined by the manufacturers.

Biopolymers. Chitosan samples were provided by Vanson, Inc., Redmond, WA, and put into solution at high concentration by the addition of weak acid solutions. After dilution of these stock solutions, the pH was adjusted back to 7.0.

Most starch and cellulose derivatives were produced "in-house" using Sigma or Aldrich reagent grade chemicals. Calcium was added to test solutions by the addition of Aldrich reagent-grade calcium nitrate stock solutions. Acid hydrolyzed cellulose microfibril suspensions were formed from cotton fiber (Whatman filter paper #4) following a procedure outlined by Revol *et al.* (*11*). Cellulose fibers were milled in a Wiley Mill to pass through a 40 mesh screen, added at 8% concentrations to ~60% sulfuric acid at 60°C, and stirred for 30 minutes. The reaction was stopped by adding excess water, and the samples were centrifuged and washed repeatedly (at least three times) until clean of salts and acid. The chemically bound sulfur from this recipe, determined previously by titration and standard elemental analysis, corresponded to a sulfonation, or surface charge, of about 10% of the surface anhydroglucose units or roughly 0.2 sulfate groups per nm^2.

Starch xanthate was produced following the procedure of Menefee and Hautala (*12*) using Midsol 50 wheat starch purchased from Midwest Grains. Starch was swollen by soaking in ~20% (w/w) aqueous NaOH. After decanting off excess NaOH, molar volumes of carbon disulfide were added with mixing, forming an orange sticky mass. After several hours of stirring, xanthates were diluted to a 2% solution (based on polysaccharide content), which presumably stopped the reaction and facilitated further dilution. The degree of substitution, ds, was estimated using standard titration methods for sulfur content. The shelf-life of xanthate is generally limited to several days, so samples were tested within 24 hours after production.

Lab-Scale Erosion Test. Lab-scale rills or furrows were created using a modified, smaller version of the soil flume tests developed at the USDA-ARS National Soil Erosion Research Laboratory (*13*). Except where noted, the soil tested was a Zacharias gravelly clay loam soil obtained from Patterson, CA, a northern California farming community 90 miles south of Sacramento. This soil was chosen because it is typical of the northern Central Valley and because PAM has been particularly effective in controlling its erosion. Soils were dried, sieved and re-moisturized to 18% (w/w) water contents and then formed into "standard" furrows which were roughly 1/100th the size of a full furrow. That is, 1500 g of moist soil was packed flat into a 2.5 x 2.5 cm well cut into a 1m long bar. A furrow with dimensions 0.63 x 0.63 cm was pressed lengthwise down the center of this soil rill, to create a miniature furrow. The furrow was set at an angle of 5° and test solutions were pumped down the furrow at standard flow rates, providing a shear rate profile similar to that achieved under field conditions. Water was collected at the lower end of the furrow and tested for solids contents, as determined by UV-Vis absorption. By comparing turbidity measurements (using a Shimadzu UV1601 Spectrophotometer) with those from a set of standard soil dispersions, the relative suspended solids of the water at the end of the furrow was determined. The "cleanliness" of this run-off water correlated to the efficacy of different polymer solutions in controlling erosion. It should be noted that the effectiveness of PAM varies with type of soil and the charge balance of the water.

Results and Discussion

Although we have screened a wide array of polymers for their ability to control erosion when added to irrigation water, this report will focus on several of the more promising biopolymer solutions. The effectiveness of any of these polymers depends on a complex interaction of (i) the soil, which is often, but not necessarily, charged, (ii) the polymer, which is often large and charged, and (iii) the water, which contains a wide array of ions. As will be discussed, the amount of available calcium in the water has a significant impact on results.

Polyacrylamide, PAM. As noted, PAM has already been proven effective in controlling irrigation-induced erosion in field applications (1-5). PAM is explored in this study for two reasons; to verify the effectiveness of the furrow test to rapidly screen polymer solutions, and to highlight some of the properties of PAM that are critical for its success. Understanding the properties of PAM will help to establish biopolymers that can mimic its erosion control benefits.

Several characteristics of PAM determine how it interacts with soil; its size, its affinity to minerals, and its conformation in water. Figures 1a and 1b outline the effect of size and charge on the efficacy of PAM in our experimental furrow test. As shown in Figure 1a, the resulting run-off water is cleaner as a function of increasing molecular weight, especially with molecular weights over 200,000. The lab-furrow results are compared with field test data from Idaho reported by Lentz *et al.* (14), which show a similar trend. Considering that the error in our data is ~3-4% of the control, it is not clear that there is significant improvement for molecular weights over 6 million. However, the field tests from Lentz *et al.* (14) and information from Cytec, a PAM supplier, suggest that results are best for molecular weights exceeding 10 million. At such high molecular weights, PAM would have a fully extended end-to-end distance of several microns and would be able to form stable flocs with multiple soil particles. Formation of large, stable soil flocs from solution is key to this process. The difference between field results and lab tests are likely attributable to the fact that the soil in the lab test has been chosen specifically because of its sensitivity to PAM treatment.

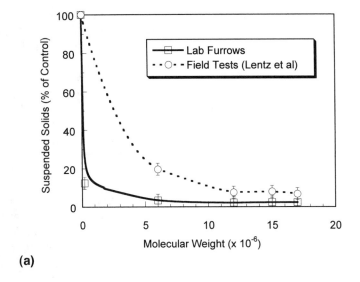

(a)

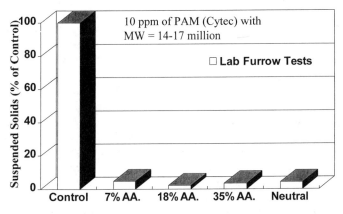

(b)

Figure 1. Lab-scale furrow tests highlighting (a) the effect of molecular weight and (b) charge type and density on the effectiveness of polyacrylamide, PAM, in controlling erosion-induced soil losses in run-off water. To promote flocculation, calcium (1.2 mM) was added to all of the solutions. Field test results from Lentz *et al.* (*14*) are included in Figure 1a for comparison.

The adsorption of PAM onto different minerals varies with the charge type and density. For clays with anionic surfaces (i.e., montmorillonite), Aly and Letey (15) reported that the increasing order of adsorption for PAM was anionic < neutral < cationic, with adsorption increasing linearly with charge density. In such a case, adsorption appears to be a simple matter of charge affinity. For whole soils with mixed mineral content, the picture is not as clear. Lentz et al. (14) reported field data suggesting that cationic PAM is not nearly as effective as anionic PAM and that neutral PAM works as well as anionic. In our lab furrow test using high molecular weight PAM, there is no significant difference between PAM-acrylic acid (AA) copolymers with different charge densities (Figure 1b). Interestingly, neutral PAM is equally effective. This highlights the importance of Van der Waals attraction during formation of soil/PAM aggregates.

The effectiveness of PAM relies on a complex interplay of polymer, ions in the water and soil. Modeling such a complex system is likely not easy. However, it appears that it is critical to form stable soil flocs during irrigation. In our selection of biopolymers that would mimic PAM, we focus on large, charged biopolymers that would form stable flocs.

Starch-Based Biopolymers. Starch-based derivatives hold promise for several reasons. The molecular weight of starch, or at least of its amylopectin component, is greater than PAM. Starch can be modified to carry a charge and to dissolve readily in water. For example, a starch prepared by "polymerization of acrylamide-acrylic acids onto starch" (1) was field tested during furrow irrigation by Lentz et al. It did not control erosion very effectively in the first irrigation, but it worked as effectively as PAM in subsequent irrigations, after it settled and dried in the furrow.

Starch and cellulose xanthates have been considered as soil stabilizers (12, 16-18) and in the removal of heavy metals from waste water (19,20). Preliminary data from the lab-scale furrow test (Figure 2) reveal that starch xanthate (degree of substitution, ds = 1.7) can effectively reduce soil run-off when it is applied at ~9 times the recommended PAM concentration. Starch xanthate is estimated to cost roughly less than $1/lb, and more likely about one-fifth the retail price of PAM (sold at about $4.00 per lb). Also, it should be noted that starch xanthate with ds = 1.7 may not have the optimal degree of substitution for this application. With further work, xanthates may prove to be a very useful alternative for PAM.

Several drawbacks of xanthate must be overcome, although they have been addressed before. The relative instability of xanthates means it has a shelf life of days or weeks. Meadows (16,17) developed strategies to extend the shelf life by removing the water, by storing at cooler temperatures, by storing in vacuum-sealed packages and by adding dehydrating agents, such as CaO. In the large-scale production of xanthates, the toxicity, flammability and odor of carbon disulfide is unappealing. This issue has been addressed in the production of rayon, but has likely led to the shrinking appeal of the viscose process in making rayon.

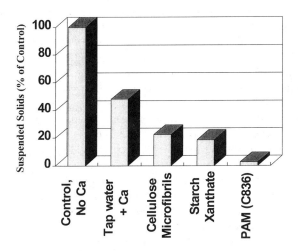

Figure 2. The effectiveness of cellulose microfibrils and starch xanthate on controlling soil run-off in a lab-scale furrow test. The control was pure tap water. For all other samples, polymer was added to tap water with 0.63 mM added calcium. The concentrations that were required for cellulose microfibrils (120 ppm) and starch xanthate (90 ppm) were higher than that used for PAM (20 ppm). The PAM had a MW of 16 million and charge density of 18% via randomly distributed acrylic acid groups.

Cellulose Microfibrils. Most of the biopolymers that were screened in this study had been previously touted in the literature as either a soil conditioner or as a flocculating agent. This is not the case for charged cellulose microfibrils, which have generally been considered for applications in the paper, textile and drug industries. Yet, microfibrils appear to possess the major attributes required for creating stable soil aggregates; i.e., a large size, an affinity to soil via a surface charge and easy dispersion in water. Cellulose microfibrils are the "basic crystalline unit" of a cellulose fiber obtained during acid hydrolysis. Microfibrils derived from cotton are generally stiff rods with a length of 0.1-0.3 microns and a diameter of 5 nm (*11,21*). These dimensions approach those of a fully extended, high molecular weight PAM. Microfibrils are readily dispersed in water because of their surface charge, which in this case is derived from the anionic sulfate esters created during hydrolysis in sulfuric acid.

Aqueous suspensions of cellulose microfibrils reduce the irrigation-induced erosion in the lab-scale furrow experiments. Concentrations of at least 100 ppm were required to exhibit any positive result, with a concentration of 120 ppm resulting in 78% reduction of solids in the run-off (see Figure 2). In contrast, PAM (Cytec Magnifloc 836A) removed 96% of solids at a concentration of 20 ppm. However, users of PAM in the field report positive results with concentrations as low as 2ppm.

Despite the fact that relatively high concentrations of cellulose were required, the charged microfibrils are still promising for several reasons. First, as with starch xanthate, the charge distribution of the microfibrils has not been optimized. A wide range of charge density and charge type can be obtained by optimizing reaction conditions. Second, the optimal microfibril size has not necessarily been chosen. It is possible to vary the size of the microfibrils by varying the source of the cellulose. For example, cotton microfibrils are significantly smaller than the fibers derived from acid hydrolysis of sugar beet. Third, it may be an advantage to use much higher concentrations of a biopolymer in the field compared to PAM. Steps are taken to minimize the use of PAM because it is a reasonably expensive, synthetic polymer, which can not be introduced into the environment at high concentrations without potential negative environmental impact. On the other hand, adding relatively high concentrations of cellulose back to soil may be a useful source of structural polysaccharide for soils that generally lack sufficient structure. Finally, using microfibrils derived from agricultural fibers presents new uses for waste agricultural products.

Chitosan. If production costs were no consideration, chitosan would be a logical alternative to PAM for controlling erosion-induced soil losses. Chitosan, which is chemically similar to cellulose with the hydroxyl in the 2-position replaced with a primary amino group, has a net positive charge at neutral or acidic solution pH values. Chitosan has already established itself as a premium-priced biodegradable flocculating agent for removing heavy metals from plant water (*22*), for reducing suspended biological matter in municipal waste (*23*) and for clarifying swimming pool water in an "environmentally friendly" manner (*24*). Lab furrow results outlined in Table I show that highly deacetylated chitosan at 20 ppm is nearly as effective as PAM in reducing erosion-induced soil losses. The major drawback of chitosan is its cost of at least $8-10/lb, which is more than twice the price of PAM. Another drawback of chitosan is its low solubility in water. It is generally recognized that chitosan with deacetylation of < 0.3 is only readily soluble at a pH of 5.5 or less. This problem has been overcome by a number of methods including that used by Vanson Inc., Redmond, WA, which recently introduced a fully water-soluble chitosan. The key to improving the cost/performance of chitosan would be to develop a wider array of inexpensive water-soluble derivatives and to investigate the use of less purified grades of chitosan. Rather than using pure, high-grade chitosan as a soil additive, it may be possible to tolerate degradable impurities in a chitosan sample.

Calcium Effects. Regardless of which polymer was investigated here, results were affected by the concentration of (exchangeable) calcium in the solution. A comparison of the two columns in Table I reveals that, with the addition of 1.2 mM of calcium, the suspended solids were reduced significantly. For the PAM solution, calcium reduced suspended solids from ~3.5 mg/mL to trace amounts. An explanation for this effect can be found in the following related observations: (i) calcium can act as an "ionic bridge" between polymer and soil charges (ii) calcium carries a multivalent charge which can replace the other shielding ions in the soil minerals and (iii) calcium changes the radius of gyration (R_g) of the PAM molecule in solution, with R_g decreasing as a function of increasing calcium.

It would be difficult to develop a mechanistic picture of soil/PAM interactions that would fully explain the relationship of these various ions. However, it is clear that calcium improves the PAM/soil adsorption (Table I, row 2). Wallace and Wallace (5) have reported similar results by combining PAM with gypsum, an inexpensive source of exchangeable calcium. They noted that gypsum and PAM mixtures would lower the required amount of PAM required to control erosion. Such synergy is probably not restricted to PAM use and may play a significant role in our formulations of biopolymers that mimic PAM.

Table I. A Comparison of Polyacrylamide, PAM, (MW ~ 16 million, 18% anionic) and Chitosan Solution in Controlling the Extent of Suspended Solid Run-Off in a Lab-Scale Furrow Test. For All Samples, Adding Calcium Nitrate at a Concentration of 1.2 mM Greatly Reduced the Suspended Solids in the Effluent

	Without Calcium	*Calcium Nitrate Added (1.2 mM)*
Control (Tap Water)	48.1 ± 5.1	1.6 ± 0.2
PAM (20 ppm)	3.4 ± 0.4	0.01 ± 0.01
Chitosan (20 ppm)	5.5 ± 1.8	0.05 ± 0.01

One of the more striking results of this study was the fact that improvements in soil run-off in the lab furrow test were achieved independent of the presence of PAM. Suspended solids are reduced by more than 85% by adding 1.2 mM of calcium to the tap water control (Table I, row 1). It should be noted that the soils tested in this study consist mainly of clay, which generally carries a charge. To explore this result further, we utilized a soil from the Davis, CA, area that was rich in clay and did not respond well to PAM. As shown in Figure 3, calcium alone significantly reduced suspended solids in the run-off from this soil, although PAM and calcium still had a greater effect than calcium alone. A similar result was also reported by Wallace and Wallace (5), who noted that gypsum or other ions providing higher electrical conductances (25) can flocculate clays. Our result confirms their findings and extends it from a simple settling test to a furrow run-off test. Note though, that this result may not be universal and likely reflects the fact that the soil is clay-rich and that the "tap water" used here does not have an "excess" of exchangeable calcium. Also, the benefit from added calcium is short term; i.e., calcium must be added continuously to the irrigation water. In contrast, PAM can be added for a short period during an initial irrigation series and still provide an effect lasting for weeks of irrigation cycles without PAM. In mechanistic terms, we can surmise that calcium "salts out" soil from the water without necessarily providing structure to the soil. There is a long-term advantage in using PAM or, perhaps, one of the biopolymers screened in our study to promote soil structure.

Conclusions

A lab-scale furrow test has been established to test different aqueous polymer formulations for their efficacy in reducing erosion losses during irrigation. After exploring the effect of size and charge of PAM on its ability to interact with soil, an array of biopolymers with similar properties was tested. Starch xanthate and acid hydrolyzed cellulose microfibrils each appeared promising, with the ability to reduce soil run-off significantly. The effective concentrations of these derivatives, though, were up to 10 times higher than for PAM, without even matching PAM's full efficacy. These polymers would not cost as much as the "high-grade" PAM employed here. Therefore, these high concentrations would not necessarily exclude these polymers on an economic basis, especially if they are optimized for this application.

Chitosan solutions are effective at controlling erosion-induced soil losses at concentrations approaching those used for PAM. Its major drawback would be its present price, with the cost of application under the conditions outlined here at least double that of PAM. The economics of chitosan may be improved by providing polymer of lower purity, thus saving production costs.

Calcium has a significant effect on the application of all of the polymers screened in this study, at least for the clay-rich soils encountered in Northern California. Increasing the exchangeable calcium levels in the water improves the effectiveness of PAM. Interestingly, calcium nitrate alone can reduce the amount of soil run-off, although it has no long-term effect. These results support suggestions by

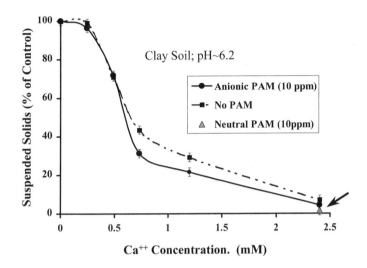

Figure 3. The effect of added calcium on suspended solids in soil run-off from a lab-scale furrow test for water with and without PAM (MW~16 million, 18% anionic). For soils rich in clay, which is charged silica, calcium alone reduced erosion-induced soil losses. Note also, that neutral PAM was at least as effective as PAM in reducing suspended solids, as indicated by the arrow in the figure.

246

Wallace and Wallace (5) that an optimal formulation will likely consist of polymer and a source of exchangeable calcium to improve the flocculation process. Our preliminary screening results discussed here introduce several biopolymers that may prove viable in this application.

Acknowledgements

The authors thank Bob Sojka for creative insights, encouragement and support, Mike McElhiney for helpful discussions and Youngla Nam for technical support. Financial support by the Washington and Idaho Wheat Commissions is gratefully acknowledged.

Literature Cited

1.   Lentz, R. D.; Shainberg, I.; Sojka, R. E.; Carter, D. L. *Soil Sci. Soc. of Am. J.* **1992**, *56*, 1926.
2.   Lentz, R. D.; Sojka, R.E. *Soil Sci.* **1994**, *158*, 274.
3.   Trout, T. J.; Sojka, R. E.; Lentz, R. D. *Trans. ASAE* **1995**, *38*, 761.
4.   Levy, G. J.; Ben-Hur, M.; Agassi, M. *Irrig. Sci.* **1991**, *12*, 55.
5.   Wallace, A.; Wallace, G. A. In *Proceedings: Managing Irrigation-Induced Erosion and Infiltration with Polyacrylamide*; Sojka, R. E.; Lentz, R. D., Eds.; Twin Falls, ID, May, **1996**.
6.   Kay-Shoemake, J. L.; Watwood, M. E. In *Proceedings: Managing Irrigation-Induced Erosion and Infiltration with Polyacrylamide*;. Sojka, R. E.; Lentz, R. D., Eds.; Twin Falls, ID, May, **1996**.
7.   Kawai, F. *Appl. Microbiol. Biotech.* **1993**, *39*, 382.
8.   Barvenik, F. W.; Sojka, R. E.; Lentz, R. D.; Andrawes, F. F.; Messner, L. S. In *Proceedings: Managing Irrigation-Induced Erosion and Infiltration with Polyacrylamide*; Sojka; R. E.; Lentz, R. D., Eds.; Twin Falls, ID, May, **1996**.
9.   Sojka, R. E. USDA-ARS Soil and Water Management Research Unit, Kimberley, ID, personal communication, **1998**.
10. Product names, company names and trademarks are included in this report to facilitate proper description of experimental methods. Use or discussion of these products does not imply endorsement by the USDA or the ARS.
11. Revol, J.-F.; Bradford, H.; Giasson, J.; Marchessault, R. H.; Gray, D. G. *Int. J. Biol. Macromol.* **1992**, *14*, 170.
12. Menefee, E.; Hautala, E. *Nature* **1978**, *275*, 530.
13. Stott, D. E; Trimnell, D.; Fanta, G. F.. In *Proceedings: Managing Irrigation-Induced Erosion and Infiltration with Polyacrylamide*; Sojka, R. E.; Lentz, R. D., Eds.; Twin Falls, ID, May, **1996**.
14. Lentz, R. D., Sojka, R. E. In *Proceedings: Managing Irrigation-Induced Erosion and Infiltration with Polyacrylamide*; Sojka, R. E.; Lentz, R. D., Eds.; Twin Falls, ID, May, **1996**.
15. Aly, S. M.; Letey, J. *Soil Sci. Soc. Am. J.* **1988**, *52*, 1453.

16. Meadows, G. W. U.S. Patent 2,761,247; **1956**.
17. Meadows, G. W. U.S. Patent 2,884,334; **1959**.
18. Swanson, C. L.; Wing; R. E.; Doane, W. M. U.S. Patent 3,947,354; **1975**.
19. Maher, G. G. U.S. Patent 4,253,970; **1981**.
20. Coltrinari, E. U.S. Patent 5,320,759; **1994**.
21. Marchessault, R. H.; Morehead, F. F.; Walter, N. M. *Nature* **1959**, *184*, 632.
22. Deans, J. R. U.S. Patent 5,336,415; **1993**.
23. Murcott, S. E.; Harleman, D. R. F. U.S. Patent 5,543,056; **1994**.
24. Nichols, E. Vanson Inc., Redmond, WA, company literature and personal communication, **1997**.
25. Shainberg, I.; Letey, J. *Hilgardia* **1984**, *52*, 57.

Chapter 17

Two Food Applications of Biopolymers: Edible Coatings Controlling Microbial Surface Spoilage and Chitosan Use to Recover Proteins from Aqueous Processing Wastes

J. A. Torres, C. Dewitt-Mireles, and V. Savant

Food Process Engineering Group, Department of Food Science and Technology, Oregon State University, Corvallis, OR 97331–6602

Edible coatings prepared using proteins and polysaccharides can help retain high preservative concentrations on food surfaces to retard microbial surface growth which limits the shelf life of many products. Coating effectiveness is affected by coating formulation, food water activity, food pH and storage temperature. Microstructure determinations by electron microscopy help interpret coating effectiveness. Chitosan, the deacetylated derivative of chitin, is a versatile molecule with potential applications in diverse fields, including waste water treatments. By electrostatic interactions of chitosan NH_3^+ groups with COO^- or SO_3^- groups in proteins or polyanions (e.g., alginate, carrageenan and pectin), chitosan can form polymeric complexes. Suspended proteins found in waste effluents, fermentation media and other industrial streams can be recovered as chitosan-protein or chitosan-polyanion-protein complexes.

Edible Coatings to Control Microbial Surface Spoilage

Edible coatings developed with a wide range of properties contribute to the stability of many processed foods (Table I). An edible coating or film is defined as a thin, continuous layer of edible material formed or placed, on or between, foods or food components to provide a barrier to mass transfer, to serve as a carrier of food ingredients and additives, or to provide mechanical protection. Polysaccharides such as cellulose, modified cellulose and starch; proteins such as whey, wheat and soy proteins, zein, collagen, gelatin, ovalbumin and serum albumin; plant and microbial polysaccharides such as agar, carrageenan, alginate, pectin, dextrin, gum ghatti, scleroglucan, pullulan, curdlan; and waxes and lipid derivatives have all been considered for food applications (1-29).

Restricting the water activity (a_w) in foods is important to control microbial growth, prevent texture degradation and minimize undesirable chemical and enzymatic

248

Table I. Applications for Edible Coatings and Films

Control of moisture migration/losses

Control of gas exchanges (O_2, CO_2, C_2H_4, etc.)

Control of oil and fat absorption and migration

Control of solute migration

Control of flavor and other volatile migration, exchange, and losses

Carrier of flavor, color, antimicrobial, and other food additives

Prevention or control of photo degradation - oxidation

Improvement of mechanical handling properties of foods

Adapted from (*37*).

reactions. Consequently, a functional property of key interest in edible films and coatings is resistance to moisture migration. Most edible films and coatings use polymers to provide structural support, a lipid to improve moisture barrier effectiveness, and a plasticizer to modify mechanical properties. Naturally occurring examples of these multicomponent systems are the glycoproteins, lipoproteins and glycolipids present in cellular membranes (*29*).

Fennema and coworkers have investigated the control of moisture migration between regions with different water activities by methyl and hydroxypropyl methyl cellulose films. Films containing stearic acid, beeswax, paraffin or hydrogenated palm oil had lower water permeabilities than low density polyethylene (*4,31-36*). Extensive research has been done also on films and coatings using proteins as the structure-forming component with added lipids and plasticizers to improve their functionalities (*37-38*).

Surface microbial stability determines the shelf-life of many products including refrigerated and intermediate moisture foods (IMF) (*10-11,37*). During storage and distribution of meats, beef, poultry and seafood, nearly all microbial growth occurs on the surface with Gram-negative bacteria being predominant (*39-47*). For example, fresh broilers in retail outlets have initial surface counts of 10^4 to 10^5 microorganisms/cm^2, and as a result can be stored refrigerated only for a few days before spoiling (*48-49*).

In IMFs, low a_w and preservatives retard or stop the growth of bacteria, molds and yeasts (*50*). Progress in the development of these IMFs and other foods can be conceptualized by representing shelf-life and organoleptic quality as a function of the fabrication parameters; i.e., production, formulation and processing steps. The concentrations of stability additives (e.g., salt or K-sorbate) and the intensity of processing steps (e.g., heating) increase product stability but generally decrease perceived quality. An example of this approach, corresponding to the development of an intermediate moisture cheese analog, illustrates the product development difficulties surrounding IMF technology (Figure 1). The search for potentially acceptable formulations was guided by three criteria: (1) texture, evaluated with an Instron; (2) taste, identified by sensory restrictions; and, (3) a minimum a_w, measured with an electric hygrometer. The limited number of acceptable combinations, represented by the shaded area, points out how difficult it would be to add considerations on product abuse. For example, storage temperature fluctuations (*51*), product transfers between facilities at different temperatures, or packaging products while still warm affect microbial stability of food surfaces (*2,8*). These situations result in surface condensation leading to localized increases in surface a_w where microbial growth occurs even if the bulk a_w is acceptable (*2,52*).

Another source of microbial instability, even for products that are heat processed, is product handling. Slicing and packaging provide many opportunities for surface recontamination (*53*). Surface counts are highly variable (*43-44*) and this is particularly important for IMFs since bacteriostatic barriers can be overcome by large localized microbial counts. Furthermore, the minimum a_w for certain microorganisms is oxygen-dependent. Under anaerobic conditions the minimum a_w for *Staphylococcus*

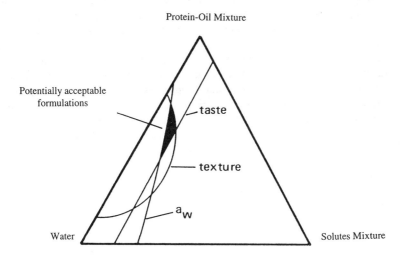

Figure 1. Product development difficulties surrounding IMF technology. Adapted from (8).

252

aureus is 0.91, whereas under aerobic conditions it is 0.86 (*54*). Product developers should be more concerned with potential outgrowth of this ubiquitous organism on food surfaces where oxygen is more readily available.

To cope with surface microbial problems, food processors treat food surfaces with approved preservatives. Potassium sorbate dips reduce viable bacteria at refrigeration and temperature abuse conditions (*48,55-59*). However, the shelf life extension achieved is limited because diffusion reduces preservative concentration on the surface where microbial spoilage must be controlled (*8,60*). Sorbic acid diffusion rates (D) can be reduced by lowering food a_w. Lowering the a_w of a model system from 1.0 to 0.88 by use of 40% w/w glycerol or 16% w/w salt reduces the apparent diffusivity at room temperature from 6.7×10^{-6} to 2.0×10^{-6} cm^2/s. At 70% glycerol, the a_w is 0.64 and D is 5×10^{-7} cm^2/s (*5*). Surface microbial stability can be further improved by retaining a higher (initial) concentration of preservative(s) on the surface and by using a coating to maintain a large concentration difference for as long as possible (*8*). This approach requires coatings to reduce preservative diffusion from food surface into food bulk (Figure 2). For example, films made from zein, a corn protein, combined with potassium sorbate show improved surface microbial stability. The effectiveness of zein films was confirmed in microbial tests using a model food system with $a_w = 0.88$ coated with zein and *S. aureus* as the challenge microorganism. The barrier property of zein films was identified as the mechanism for stability improvement (*2-3*). The diffusion of sorbic acid in zein films was found to be $3-7 \times 10^{-9}$ cm^2/s, i.e., about 100-1,000 fold slower than in foods. The film is very water-resistant but may add off-flavors (*6*) and is expensive.

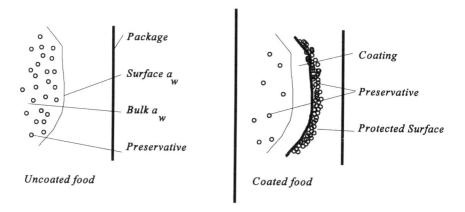

Figure 2. Preservative diffusion from food surface into food bulk controlled by edible coatings. Adapted from (*8*).

Permeability Measurements

Permeability can be determined using a cell consisting of two mechanically agitated chambers separated by the film to be tested (2,8). The upper chamber is filled with aqueous glycerol or pure water. The lower chamber contains the same solution with 2.5% w/v potassium sorbate. The cell is placed in a controlled temperature chamber. When mounted on the permeability cell the top side (air drying side) of the film faces the high K-sorbate concentration. The use of glycerol solutions is recommended because it allows the inclusion of the food a_w effect on the coating permeability. Samples are taken from the upper chamber and the K-sorbate concentration is measured spectrophotometrically at 255 nm. Films should be inspected before and after every test to assure that results are not affected by cracks or other types of film failures. The permeability test is not a gentle procedure and films are subjected to the mechanical abuse of stirrers and compression between the two permeability cell chambers.

Permeability coefficients are calculated by plotting preservative transferred through the film as a function of time (2,8). After a time lag, the slope of the linear relationship obtained is the steady-state rate of K-sorbate transferred through the film (61,62).

$$K = \frac{Fl}{c} \qquad (1)$$

where K is the apparent permeability coefficient, F is the amount of K-sorbate permeated per unit time, l is the thickness of the film, and c is the K-sorbate concentration in the lower chamber. The determination is confirmed by the following relation between lag time and apparent permeability constant (Equation 4.26 in 61,62):

$$L = \frac{l^2}{6K} \qquad (2)$$

where L is the intercept on the time axis by extrapolation of the steady-state rate of K-sorbate transfer through the film. For example, in tests at 0.80 a_w for a polysaccharide film with thickness 66 μm the intercept on the time axis was 6 min and the calculated value using Equation (2) was 5.86 min. A film with the same composition and 63 μm thickness tested at 0.65 a_w had an observed lag time of 8 hr and a calculated value of 8.2 hr (19). Similar determination and confirmation procedures can be used for other permeating molecules and films.

Factors Controlling Potassium Sorbate Permeability

Coating effectiveness is affected by the polysaccharides used in coating formulations, storage temperature, lipid component, food water activity, and food pH.

Temperature Effects. The permeability phenomenon is a combination of two physical processes. There are sorption and desorption processes on both sides of the membrane, which depend on the solubility of the diffusing molecule in the film and the nature of the adhesive forces at the interface (*63*). In addition, the permeate must diffuse in the film. In most cases, the latter process is the controlling step and explains why permeability rates follow an Arrhenius activation energy model (*30,63-65*):

$$K = K_o \, e^{\frac{-E_a}{RT}} \qquad (3)$$

where, K_o is a frequency constant, E_a is the permeation activation energy, R is the universal gas constant, and T is the absolute temperature. For example, values for potassium sorbate permeability through polysaccharide films were measured at 5, 24, 32, and 40°C (Table II) and used to obtain Arrhenius plots. The lack of breaking points in the Arrhenius plots (Figure 3) indicates that no film morphological changes occur in the 5 to 40°C range. In addition, no significant differences ($\alpha = 0.05$) exist between the slopes for films made from chitosan, methyl cellulose, hydroxypropyl methyl cellulose, and a mixture of the latter two, when the permeability was measured using a glycerol solution. However, the slope difference between the chitosan film in water versus all other films in aqueous glycerol was highly significant, with a 45% reduction in activation energy (Table III).

The observation that permeability values follow the Arrhenius model and that the activation energy is affected by the solvent embedding the film suggests that the diffusion process in the film occurs through the aqueous phase. Consequently, the performance of edible coatings controlling surface preservative concentration will depend strongly on the food a_w. However, K_o values are affected by the nature of the polysaccharide.

Another approach to estimate the effect of temperature on permeability rate is by use of the following expression (*66-67*):

$$\frac{K\mu}{T} = \psi \qquad (4)$$

where μ is the solvent viscosity at temperature T and ψ is a constant. This expression is based on the Stokes-Einstein equation for the diffusion of a molecule in a medium of known viscosity (*5*). This equation should be used with caution when the solution viscosity is high because it can overestimate the temperature effect on the diffusion constant (*66*).

Fatty Acid Effects. Permeability decreases as fatty acid concentration increases (Table IV). Film casting difficulties interrupt the tendency towards lower permeability values as the fatty acid concentration increases because of the need for MC or HPMC structural support. Commercial users should explore the highest fatty acid

Table II. Potassium Sorbate Permeability Through Polysaccharide Films Evaluated at Various Temperatures

40°C			32°C			24°C			5°C		
$Kx10^8$	l	$\psi x10^{11}$	$Kx10^8$	l	$\psi x10^{11}$	$Kx10^8$	l	$\psi x10^{11}$	$Kx10^8$	l	$\psi x10^{11}$
Chitosan film/glycerol solution											
13.7	21	285[q]	9.9	21	283[q]	8.6	24	320[pq]	3.6	20	355[p]
17.1	25		11.4	23		8.5	24		3.5	27	
13.7	20		11.4	23		8.8	23		3.6	22	
				Average $\psi = (311 \pm 34)x10^{-11}$							
Hydroxypropyl methylcellulose (HPMC)/glycerol solution											
13.9	98	269[r]	10.6	95	275[r]	8.1	98	308[q]	3.6	96	367[p]
13.8	92		11.1	90		8.5	120		3.8	92	
14.4	110		10.2	100		8.3	95		3.6	95	
				Average $\psi = (305 \pm 45)x10^{-11}$							
Methylcellulose (MC)/glycerol solution											
6.2	99	137[p]	5.0	113	131[p]	3.2	95	125[p]	1.4	140	144[p]
6.5	100		5.7	112		3.3	103		1.4	110	
7.5	104		4.6	90		3.6	120		1.5	120	
				Average $\psi = (134 \pm 8)x10^{-11}$							
HPMC+MC/glycerol solution											
10.0	95	198[q]	8.5	98	214[q]	5.9	90	216[q]	2.6	114	244[p]
10.5	110		8.1	110		5.6	95		2.2	115	
10.4	100		8.2	104		6.0	85		2.5	111	
				Average $\psi = (218 \pm 19)x10^{-11}$							
Chitosan film/water											
95.7	28	200[p]	77.7	28	208[p]	62.5	26	200[p]	39.0	25	208[p]
83.7	24		75.1	26		63.7	24		39.9	29	
82.2	23		76.8	27		61.8	25		40.0	26	
				Average $\psi = (204 \pm 5)x10^{-11}$							

K is permeability constant in $(mg/s \ cm^2)(cm)/(mg/cm^3)$, l is film thickness in μm, ψ is evaluated with Eq. (4). ψ values with the same superindex letter are not significantly different ($\alpha = 0.05$). All tests were conducted using a glycerol solution or water.

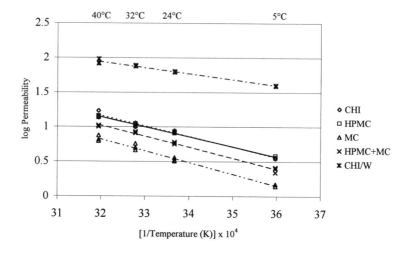

Figure 3.   Arrhenius plots for potassium sorbate permeability through polysaccharide films. Adapted from (*10*).

Table III. Activation Energy for K-Sorbate Permeation Through Polysaccharide Films

Film	E_a, Kcal/g-mole
Chitosan	6.98
Hydroxypropyl methylcellulose (HPMC)	6.62
Methylcellulose (MC)	7.71
HPMC + MC	7.27
Average	7.15 ± .46
Chitosan/water	3.95

Adapted from (*10*).

Table IV. Potassium Sorbate Permeability at 24 °C Through Polysaccharide Films with Various Fatty Acid Content

Ratio	Lauric		Palmitic		Stearic		Arachidic	
	$K\times10^{-9}$	l	$K\times10^{-9}$	l	$K\times10^{-9}$	l	$K\times10^{-9}$	l
				Methylcellulose (MC)				
45:0	44	28						
	47	29						
45:1	64	24	45	30	41	33	21	30
	63	28	44	25	42	34	24	34
45:5	53	26	37	29	36	34	15	26
	48	30	34	31	35	33	13	24
45:10	43	26	14	22	13	31	8.0	30
	41	28	14	27	11	33	7.4	31
45:15	27	34	12	34	9.4	38	4.9	42
	30	28	12	40	8.4	42	4.8	49
45:20	22	30	4.4	40	8.4	40	--	--
	21	32	5.4	36	12	30	--	—
			Hydroxypropyl methylcellulose (HPMC)					
45:0	87	29						
	78	25						
45:1	86	21	65	26	58	24	51	21
	84	21	66	29	54	21	43	22
45:5	68	21	46	28	40	27	32	20
	72	22	45	24	39	24	32	22
45:10	56	20	31	25	24	20	17	27
	50	24	31	28	26	23	19	25
45:15	43	28	20	33	16	23	9.5	29
	48	30	21	37	15	22	9.2	36
45:20	28	25	6.9	24	25	25	--	--
	28	22	5.4	27	24	30	--	

K is permeability constant in (mg/sec cm^2)(cm)/(mg/mL), l is film thickness in μm, ratio is parts of (MC or HPMC):(fatty acid). Adapted from (10).

concentrations possible in their applications. Permeability also decreases for films with longer chains for these saturated fatty acids. At all fatty acid concentrations the lowest permeability values were obtained for arachidic acid (Table IV). This is consistent with published data on the permeability of synthetic lecithin liposomes. Increasing the lecithin fatty acid chain length decreases the permeability of glycerol and erythritol through these artificial membranes (68).

Unsaturation in the fatty acids has been reported to affect membrane permeability. The permeability of liposomes and intact *Mycoplasma laidlawii* cells changes with the geometrical configuration and the number of double bonds in fatty acids (30). In synthetic liposomes, the rate of glycerol permeation increases in the order elaidic < oleic < linoleic. Rates of potassium permeability through polysaccharide films with fatty acids are consistent with these observations and, therefore, the use of films containing unsaturated fatty acids is discouraged. However, conformation had no effect (Table V).

Food a_w Effects on Permeability. Potassium sorbate diffused per unit area as a function of time and a_w (0.65, 0.70, 0.75 and 0.80) is shown in Figure 4 for methylcellulose films with an MC:(fatty acid) = 45:15 ratio. Average values of triplicate trials at constant temperature were not calculated because film thickness varied somewhat among replicates (55 - 66 μm). The K-sorbate permeation rate decreases with a_w, which is in agreement with potassium sorbate diffusion studies in model food systems (5) because less solvent is available as a diffusion medium. In addition, film hydration depends on the water activity of the food being coated and also affects the permeation rate (10).

Food pH Effects on Permeability. At constant a_w (0.80) and temperature (24°C), the sorbic acid permeation rate decreases when pH increases (Figure 5). In coated foods with pH near neutral, the surface concentration of potassium sorbate will remain larger for longer times. This is favorable because higher concentrations are needed as food pH approaches neutrality where potassium sorbate and many similar preservatives are less effective.

Control of Surface pH by Edible Coatings

The Donnan model for semipermeable membranes (69) can be used to analyze differences between coating and food pH. This model was developed to describe pH differentials between two solutions created by a membrane permeable to low molecular weight electrolytes but impermeable to charged macromolecules (70-75). The coating and the food represent side 1 and 2, respectively (Figure 6). Entrapment of the charged macromolecule in the coating is analogous to the role of the membrane in the Donnan model. Water and other small solutes, particularly electrolytes (e.g., Na^+ Cl^-), move freely between food bulk and food surface while the charged macromolecule; e.g., carrageenan, remains in the coating. This situation can be schematized as indicated in

**Table V. Effect of Double Bonds on
Potassium Sorbate Permeability at 24°C**

Film	$K \times 10^{-9}$	l
MC:C18 (stearic)	13	31
	11	33
MC:C18:1, trans (elaidic)	31	35
	33	41
MC:C18:1, cis (oleic)	36	32
	37	29

Methylcellulose (MC) films with an MC:(fatty acid) = 45:15 ratio, permeability K in (mg/sec cm²) (cm)/(mg/mL) and film thickness l in μm.

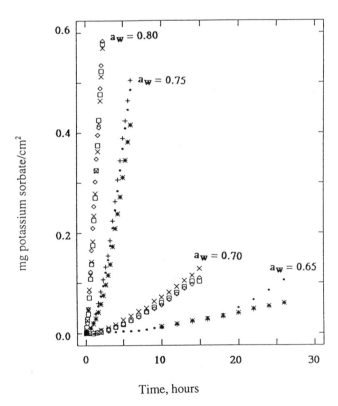

Figure 4. Sorbic acid permeated per unit area as a function of time and a_w at constant temperature (24°C) for methylcellulose films with palmitic acid (3:1 weight ratio). Adapted from (*19*).

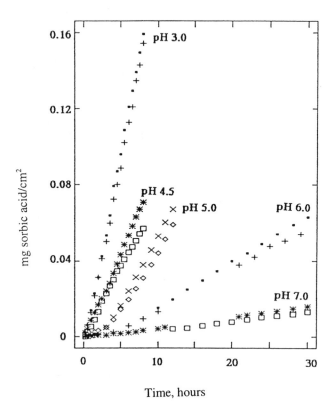

Figure 5. Sorbic acid permeated per unit area as a function of time and pH at constant a_w (0.80) and temperature (24°C) for methylcellulose films with palmitic acid (3:1 weight ratio). Adapted from (*19*).

M

Side 1	molal concentration		Side 2	molal concentration
M^+	$Kw/y + n + zp - y$			$Kw/x + m - x$
X^-	n			m
H^+	y			x
OH^-	Kw/y			Kw/x
P^{-z}	p			-----

M'

Figure 6. Schematic representation of the application of the Donnan model as applied to a charged macromolecule forming part of a food coating. Adapted from (52).

Figure 6, where M-M' represents a hypothetical membrane, M^+ and X^- are electrolytes, and P^{-z} is the charged macromolecule. Using assumed molar concentration values n, m, y, x, and p, and the equilibrium constant for water (K_w), an expression for the distribution of all permeable solutes in terms of charges and concentrations is derived from electroneutrality considerations (52).

$$\frac{\frac{K_w}{y} + n + zp}{\frac{K_w}{x} + p} = \frac{y}{x} = \lambda \qquad (5)$$

$$\lambda = \sqrt{1 + \frac{z\,p}{n}} \sim 1 \qquad (6)$$

The electrolyte concentration (n) is much larger than the product of the number of charges (z) multiplied by the molal concentration of the charged macromolecule (p). Therefore, the ratio of the proton concentration in the coating to that in the food, λ (Equation 5), approaches 1 in most coated foods (Equation 6). Since, it can be

$$\log \lambda \;=\; \log y \;-\; \log x \;=\; pH \,(food) \;-\; pH \,(coating) \qquad (7)$$

concluded that significant pH differences between food bulk and coated food surface can only be achieved in foods with low electrolyte concentrations.

Examples of Edible Polysaccharide Coatings

The permeability values reported for methylcellulose (MC) and hydroxypropyl methylcellulose (HPMC) films with fatty acids at various weight ratios show that they could increase the surface microbial stability of foods with a_w in the IMF range and pH values 3-7. Longer surface retention of sorbic acid is possible at high pH, lower a_w and lower temperature. Increased retention of sorbic acid at higher pH counteracts the lower effectiveness of sorbic acid at these pH values. A serious limitation of the research on edible coatings and films is that their preparations under laboratory conditions are not realistic simulations of practical applications. An examination of the conditions used to prepare polysaccharide films with added lipids shows that the times and temperatures will be difficult to use by the food industry. Alterations to laboratory preparation methods (e.g., *8,10-12,17-19*) will induce structural changes (e.g., changes in lipid distribution in the polysaccharide matrix) and yield coatings with different functional properties.

Single Layer Films. Film preparation methods are adapted from those described first by Fennema (*4,31-36*). Cellulose ether (MC or HPMC, 4.5 g) is placed in a 400 mL beaker and mixed with 50 mL hot water (ca. 90°C) and stirred until all particles are thoroughly melted and a uniform suspension is obtained. While continuously stirring, 100 mL ethanol is added, then the fatty acids are added at MC (or HPMC): lipid ratios of 45:1, 45:5, 45:10, 45:15 or 45:20. While stirring and heating, 0.5 g PEG is also added. The heating rate is adjusted so that, at the end of the 15 min total preparation time, the solution temperature is about 84-85°C. Suspensions with the desired MC (or HPMC): fatty acid ratio are spread thinly and uniformly while still hot (80-85°C) on glass plates (20 x 20 cm) by use of a thin-layer chromatography (TLC) applicator. A warm film spreader and plate facilitate formation of the film. The spreader is set to 0.75-0.85 mm and then slowly and steadily pulled across the plates. They are left at room temperature for 4 min to allow for lipid orientation within the wet film and then placed in an oven at 80-85°C for 15 min. After drying and cooling, and removal from the plates, most films range from 20 to 30 μm thickness. Although the thickness variation between film types is large, the variation within a film type is less than 10%. The lipid orientation of the films can be examined by scanning electron microscopy, which shows migration of the lipids to the film surface (Figures 7 and 8).

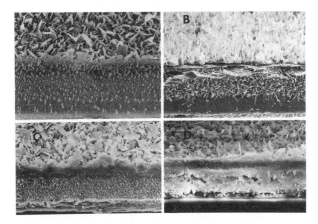

Figure 7. Electron micrographs of cross sections of edible coatings. Reproduced with permission from (*11*). (a) single layer of methyl cellulose (MC) with palmitic acid (3:1 weight ratio). (b) double layer of pure hydroxypropyl methyl cellulose (HPMC) overlaid with HPMC with arachidic acid (3:1 weight ratio). (c) single layer of methyl cellulose (MC) with stearic acid (3:1 weight ratio). (d) double layer of pure hydroxypropyl methyl cellulose (HPMC) overlaid with HPMC with stearic acid (3:1 weight ratio).

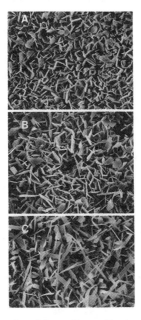

Figure 8. Electron micrographs of the surface of hydroxypropyl methyl cellulose (HPMC) coatings containing fatty acids (3:1 weight ratio). Reproduced with permission from (*11*). (a) palmitic acid; (b) stearic acid; and (c) arachidic acid.

Double Layer Films. The technique for single layer films is also used for the preparation of double layer films. The base layer contains no fatty acids while the second layer is cast with a MC or HPMC: fatty acid ratio of 45:20. The first layer is allowed to dry before applying the second layer which is done without changing the thickness setting (0.75 mm) of the thin-layer chromatography spreader. The objective of this film preparation method is to determine whether improved films can be obtained by reducing the influence of the glass plate on film casting. It should be noted that the overall permeability of these multi-layer films is controlled by the permeability of the second layer which has been shown to have much lower permeability values (10-12). No major differences in permeability have been noted on the surface of double layer films. The most significant observation from electron microscopy studies is that each layer has the same thickness and that no layer mixing occurs.

Coated Films. A base layer containing no fatty acids is first prepared as in (11,31) and then dried and peeled off the plate. It is then carefully replaced on the glass plate making sure that no air bubbles form between the plate and the film. A hot solution of lipids or edible wax (ca. 80°C) is then spread on the base layer by use of the same warm spreader set at a thickness of 0.25 mm.

MC and HPMC films coated with hydrogenated palm oil and a palmitic-stearic mixture (50% w/w) resulted in brittle surfaces that could not be tested in the permeability cell. HPMC films coated with beeswax could be tested and resulted in films with extremely low permeability values. Values reported in Table VI should only be considered an upper boundary estimate. The permeability cell is an inconvenient procedure to evaluate the barrier properties of coatings with permeability values $<\sim 10^{-10}$ (mg/s cm^2)(cm)(mg/cm^3).

Chitin and Chitosan Films. Chitin is the rigid structural component of insect and crustacean exoskeletons, and also lends rigidity to fungal cell walls. It is a structural analog of cellulose (1,4-β-D-glucopyranan) in which glucose is replaced by a 2-deoxy-2-acetamido-D-glucose; i.e., N-acetyl-β-D-glucosamine (Figure 9). Like cellulose, chitin is a white solid insoluble in water, many dilute acids and most organic solvents (85-86).

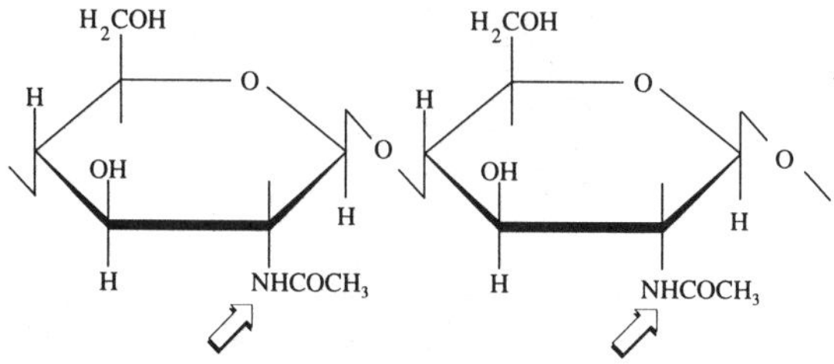

Figure 9. Chitin showing the characteristic acetoamido group - NHCOCH$_3$ on the second ring position.

Table VI. Effect of Composition and Temperature on the Permeability of Various Multilayer Films

Film	$K \times 10^9$	
Hydroxypropyl methylcellulose + beeswax at 24°C	0.045	0.058
Hydroxypropyl methylcellulose + C18-C16 mixture at 24°C	4.2	3.4
Same at 5°C	1.9	1.3

Adapted from (11).

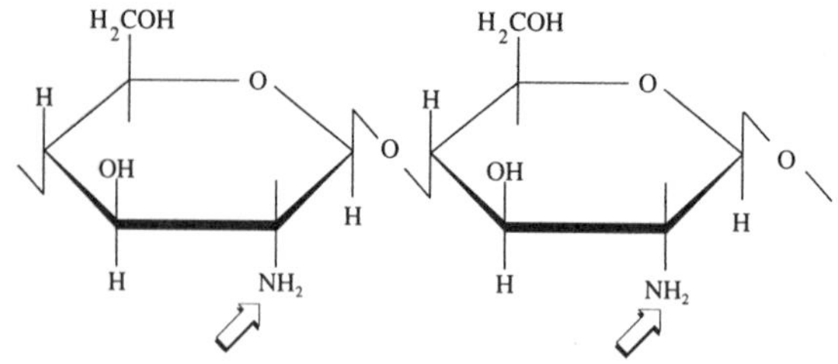

Figure 10. Chitosan showing the amino group formed by chemical or enzymatic deacetylation of chitin.

Use of microcrystalline chitin films in edible food coatings provides high temperature stability and low oxygen permeability (76). Chitosan, the deacetylated derivative of chitin (Figure 10), has received limited attention for application in coatings because it is not approved for food use in the United States and many other countries. The use of chitosan in coatings could be very effective because it inhibits the growth of a wide variety of fungi (77-79). Evans and Kent (80) showed that chitosan agglutinates a wide variety of mammalian cell types and Leube and Stossel (81) demonstrated chitosan's agglutination of sheep erythrocytes. The nature of chitosan binding to bacterial surfaces requires further examination. In a proposed mechanism, chitosan interacts with macromolecules at the cell surface and alters cell permeability (82-83). UV-absorption studies indicate that chitosan causes considerable leakage of proteinaceous material from *Pythium paroecandrum* at pH 5.8 (81). Polycation-induced leakage from plant cells is strongly inhibited by divalent cations (82) and the inhibition is competitive (84). Leakage in plant cells can be reduced by polygalacturonate presumably because of the protective formation of polyanion-polycation complexes (82).

Chitosan Use to Recover Proteins from Aqueous Processing Wastes

Chitosan is a highly versatile molecule with potential applications in vastly diverse fields ranging from waste management and medicine to food processing and biotechnology (87). Chitosan toxicity studies with animal models have shown no physiological effects (88). Chitosan-protein complexes containing up to 5% chitosan and fed to rats for 6 weeks resulted in insignificant effects on growth rate, blood chemistry, and liver physiology. This level is 10-20 times the levels expected in animal feedings (89) and may help win wide approval for chitosan use in foods. Chitosan reduces serum cholesterol and could compete advantageously with the synthetic drug cholestyramine, which has undesirable side effects. The growth rate of *Bifidobacteria*, present in the gut flora of chicken and in the intestines of infants, increases in the presence of chitin (88,90). This organism synthesizes an intracellular lactase and could help people unable to metabolize lactose. In animal feed it may improve the utilization of whey protein fractions containing lactose. The fat uptake by chitin, microcrystalline chitin, and chitosan ranges from 170 to 215% (*91*).

Chitosan Applications

Chitosan has been used effectively as an immobilizing agent for whole cells (*92-93*) and also in the recovery of plant metabolites (*94*). Immobilization of food processing enzymes such as glucose isomerase, lactase, amylases and proteases on chitin and chitosan has been successful (*94-97*). Chitosan, with its free amino groups, can function as a ligand-exchanger to recover amino acids. Ligand-exchangers using chitosan in the copper and amino copper forms are particularly effective for the recovery and separation of aspartic acid, glutamic acid, tryptophan and cysteine (*98*). Affinity-precipitation studies using chitosan as a ligand carrier indicate that this procedure is an efficient way of utilizing affinity interactions in free solution and recovering the target molecule simply by precipitation (*99*). For example, trypsin is almost completely removed from solution by chitosan with soybean trypsin inhibitor as the ligand.

Proteins have been adsorbed onto highly porous ion exchangers and adsorbents made from chitosan (*100*). The exchanger developed was a quaternary ammonium crosslinked chitosan (Chitopearl 2503, a hard gel which is not compressed by pressure in a column) and was compared to diethylaminoethyl (DEAE) Sepharose Fast Flow (hard gel) for adsorption of bovine serum albumin (BSA). Equilibrium isotherms generated for the two exchangers showed a saturation capacity of BSA on Chitopearl 2503 1.3 to 2.2 times larger than on DEAE Sepharose Fast Flow. BSA is adsorbed on the quaternary ammonium group by electrostatic attraction and on the primary amino group by the protonation reaction. At pH>pI, the electrostatic interaction is controlling while at pH = pI, the electrostatic attraction is negligibly small and BSA is adsorbed on the primary amino group by the protonation reaction (Equation 8). These ion exchangers have low manufacturing cost compared to other resins due to the abundance of chitosan.

$$R-NH_2 \ + \ protein-COOH \ \leftrightharpoons \ R-NH_3-\ ^-OOC-protein \qquad (8)$$

Chitosan is a strong positively charged molecule and acts as a chelating agent for various heavy metals (*101*). Its ability to form complexes with metals and radioisotopes makes it an ideal candidate in waste water treatment. Chitosan might have been tested to precipitate radioactive heavy metals present in milk after the Chernobyl accident (*102*). Many toxic substances such as pesticides and commercial dyes can also be removed from soils and water streams using chitosan (*103*). Such techniques as precipitation, ion exchange and electrochemical treatments have higher capital cost and are sometimes less effective, particularly when the metal concentrations are low (*104*).

Important chitosan applications have been derived from its ability to form insoluble complexes with a variety of polyanionic agents; i.e., alginates, pectins, carrageenans, sodium carboxymethyl cellulose, natural tannins, lignosulfonates and acidic glycosaminoglycans (*105*). Chitosan coagulates suspended solids due to its high amino group density, one per sugar residue, each of which may bind to groups on the suspended particulate matter. Acid-soluble crabshell chitosan and water-soluble chitosan salt have been equally effective as fining agents for apple or carrot juices (*86,106*). A one step chitosan application is as effective as the more cumbersome conventional silica sol/gelatin/bentonite treatment. A zero turbidity is reached with 0.8 kg chitosan/m^3 of apple juice.

Chitosan Applications in Waste Treatments

Chitosan shows much promise to replace the many synthetic polymers used to recover suspended food processing wastes generated by shrimp, poultry, dairy and meat processors (*107*). The suspended solids recovered as a protein-chitosan complex precipitate can be used as diet supplements in animal and aquaculture feed or further processed to recover the protein fraction (*91*). Studies conducted on the use of chitosan for protein recovery have shown that chitosan is an effective coagulating agent for suspended wastes (Table VII) (*107-109*).

Chitosan Recovery of Dairy Proteins. Increases in cheese production now require the processing of over 150 millions tons of liquid whey (*110*). The many utilization possibilities include fractionation into valuable food ingredients, production of single cell proteins and ethanol or methane, and its use in protein feeds. Unfortunately, approximately half of the world whey production goes untreated (*111*). Whey disposal has a large environmental impact because of its 30,000-50,000 ppm BOD and 60,000-80,000 ppm COD (*112*). Not only is this environmentally degrading but also a nutritional and economic waste. Whey is a rich nutritional source containing lactose (4.5-5%w/v), soluble proteins (0.6-0.8%w/v), lactic acid (0.05%w/v), citric acid, urea, uric acid and the B-group vitamins. The enforcement of stricter legislative requirements (Sec. 204 of Federal Water Pollution Control Act Amendment of 1972 [P.L.92-500]) makes it necessary that whey be converted to useful products for food, agricultural and other industries. Proteins lost in whey represent around 20% of the milk proteins (*113*). The most abundant are β-lactoglobulin (50%), α-lactalbumin (12%), immunoglobulin (10%), serum albumin (5%) and proteose peptone (0.23%). The whey proteins are rich in essential amino acids, especially lysine and have a higher protein efficiency ratio (PER = 3.4) as compared to standard casein (PER = 2.8) (*111*). They are also richer in sulfur-containing amino acids as compared to whole milk proteins.

Whey turbidity reduction by chitosan was superior to ten synthetic polyelectrolytes and resulted in the lowest value turbidity value observed (Table VIII). The protein content of the coagulated solids (Table IX) is comparable to the 80%

Table VII. Chitosan Applications in the Recovery of Suspended Proteins

Protein Source	Chitosan, (mg/L)	pH	Crude Protein of Coagulated Solids (% Dry Matter)
Cheese processing	2.5 - 15	6.0	78
Shrimp processing	60 - 360	5.5 - 6.0	65% protein recovery
Poultry processing	6 - 30	6.4 - 6.7	34 - 68
Meat processing	5 - 30	6.0 - 7.3	41
Mussel processing	40	4.5	38
Crawfish processing	150	6.0	27

Table VIII. Whey Turbidity Reduction by Chitosan and Various Synthetic Polyelectrolytes at pH 6.0

Coagulating agent[a] (30 mg/liter)	Turbidity
None	510
Chitosan (+)	150
Betz 1160 (+)	450
Betz 1190 (+)	550
Natron 88 (+)	200
Natron 6082 (+)	220
Wt-2640 (+)	530
Wt-2660 (+)	350
Wt-2263 (+)	510
Wt-2870 (+)	340
Atlasep 1N (N)	500
Atlasep 105c (+)	510

[a] + and N indicate net positive charge or neutral charge on polymer.

Table IX. Proximate Solid Analyses of Initial Cheese Whey and After Coagulation by 2.15% Chitosan

Component	Raw (%)	Coagulated (%)
Fat	1.8	0.15
Ash	9.3	9.5
Crude protein	76.4	72.3
Biuret protein	80.8	73.8
Lactose	6.6	6.0
Moisture	3.2	6.8

protein solids obtained by drying an ultrafiltration fraction. Small amounts of a polymeric coagulating agent have no adverse effects on the nutritional values of recovered proteins (*107*).

The traditional addition of an anionic polymer to dairy wastewater after acidification causes suspended particles to aggregate into larger units, which are then removed by flotation, filtration or sedimentation. The particles are essentially composed of protein carrying an outward positive charge which is coated with negatively charged polymer molecules. This is the situation when acidification takes place and the pH is below the isoelectric point (pI) for the protein (*114*). Carboxy methyl cellulose (CMC) is commonly used as the anionic polymer after reducing wastewater pH to 4.2. If the pH is > pI, the polymer has to have positive charges to give the same result. Chitosan is an example of a cationic polysaccharide that can achieve results similar to CMC even at pH 5.3. In the CMC method, sulphuric acid is used to reduce wastewater pH to 4.2 and after coagulation pH is increased to neutral levels with caustic soda. The reduction of pH from 4.5 to 4.0 requires an amount of acid equal to that required for the reduction from 7.5 to 4.5. This indicates that at pH > 4.5 it is possible to reduce by 50% the chemicals required for acidification and neutralization (*114*).

The Potential of Chitosan-Polyanion Complexes

Although chitosan can remove suspended solids (Figure 11), its cationic nature limits it to the recovery of negatively charged particles. Complexing chitosan with polyanions results in the formation of larger, potentially more effective flocculating molecules (Figure 12). Chitosan reacts with polyanions such as alginate, carrageenan and pectin by electrostatic interactions between COO^- or SO_3^- and NH_3^+ (*115*). In the 3-5 pH range, most chitosan amine groups are in the NH_3^+ form, while the polyanion reacting groups are in the -COOH (or $-SO_3H$) form, which suggests the following complex formation mechanism:

$$-NH_3^+ \ + \ -COOH \ \rightarrow \ -NH_3^+-COO^- \ + \ H^+ \quad (9)$$

At pH 6, most of the amine groups are in the NH_2 form, while most of the polyanion is in the COO^- (or SO_3^-) form which suggests the following mechanism:

$$-NH_2 \ + \ -COO^- \overset{+H^+}{\rightarrow} \ -NH_3^+-COO^- \quad (10)$$

Equation (9) indicates that complex formation at low initial pH lowers supernatant pH while Equation (10) suggests the opposite behavior at higher initial pH values. These pH-change predictions have been confirmed experimentally for chitosan complexing

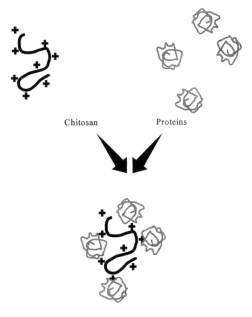

Chitosan   Proteins

Chitosan-Protein Complex

Figure 11.   Schematic representation of chitosan-protein complexes

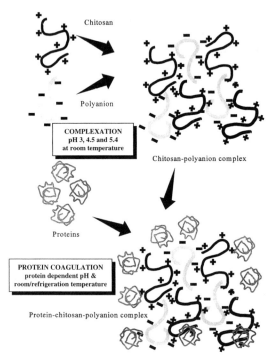

Figure 12.   Schematic representation of chitosan-polyanion-protein complexes

with polyacrylic acid, alginate, carrageenan and pectin. The electrostatic interaction was also confirmed by FTIR analyses (*105,115-117*). Interestingly, the amount of complex formed was independent of ionic strength (Figure 13). This finding has practical value since ionic concentrations in industrial waste streams vary widely.

Chitosan-polyanion complexes possess *both* positive and negative charges and could allow a more effective adsorption of proteins and other suspended solids. These complexes can be engineered according to the needs of a particular processing plant and the particulates to be recovered. We can express the mixing ratio (MR) of the reaction and the binding ratio (BR) of the complex formed as,

$$MR \quad or \quad BR \quad = \quad \frac{A}{A + B} \qquad (11)$$

where A = moles of chitosan monomer and B = moles of polyanion monomer in the reaction mixture or in the complex formed, respectively. An almost linear relationship is obtained by plotting BR as a function of MR confirming the prediction that complexes can be formed with different charge ratios (Figure 14).

Although chitosan shows much promise in many applications, most developments have remained at the laboratory level. Unfortunately successful large scale commercial applications are missing. However, the financial risk of investing in chitosan technology has tremendous reward potential, especially in the market for recovering wastes using other wasted resources: chitin and chitosan.

Acknowledgments

This work was partially supported by grant no. NA 85AA-D-SG95 (project no. E/ISG-6) from the National Oceanic and Atmospheric Administration to the Oregon State University Sea Grant College Program, by funds from the Oregon State University Agricultural Research Foundation (project 3841) and from appropriations made by the Oregon State Legislature.

Literature Cited

1. Peil, A.; Barrett, F.; Rha, C.; Langer, R. *J. Food Sci.* **1982,** *47*, 260.
2. Torres, J. A.; Motoki, M.; Karel, M. *J. Food Proc. Pres.* **1985,** *9*, 75.
3. Torres, J. A.; Karel, M. *J. Food Proc. Pres.* **1985,** *9*, 107.
4. Kester, J. J.; Fennema, O. R. *Food Technol.* **1986,** *40*, 47.
5. Guilbert, S.; Giannakopoulos, A.; Cheftel, J. C. In *Properties of Water in Foods;* Simatos, D.; Multon, J. L., Eds.; Martinus Nijhoff Publishers: Dordrecht, Netherlands, **1985**; pp. 343.

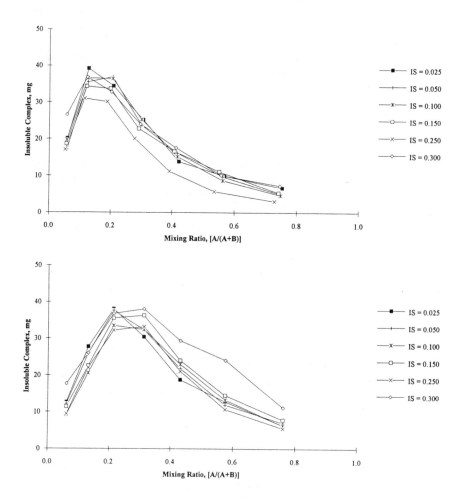

Figure 13.   Complex formation as a function of mixing ratio, pH and ionic
strength for various polyanions. Adapted from (*115*).
(a, b, c, d) Chitosan-polyacrylic acid at pH = 3, 4, 5, and 6
(e) Chitosan-alginate; (f) Chitosan-pectin; (g) Chitosan-
carrageenan

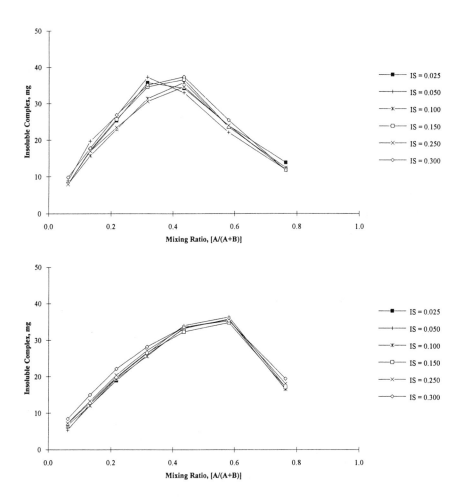

Figure 13. *Continued.*

Continued on next page.

276

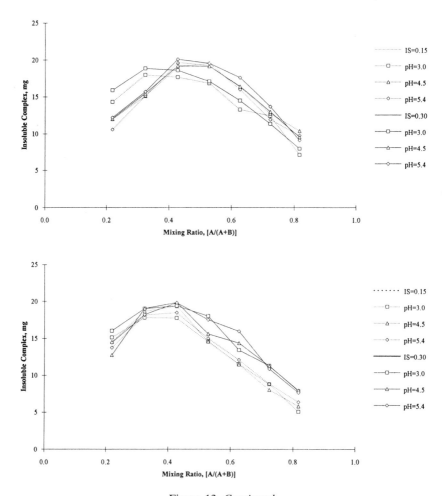

Figure 13. *Continued.*

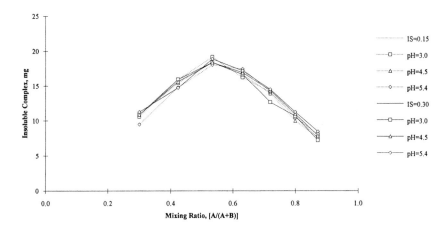

Figure 13. *Continued.*

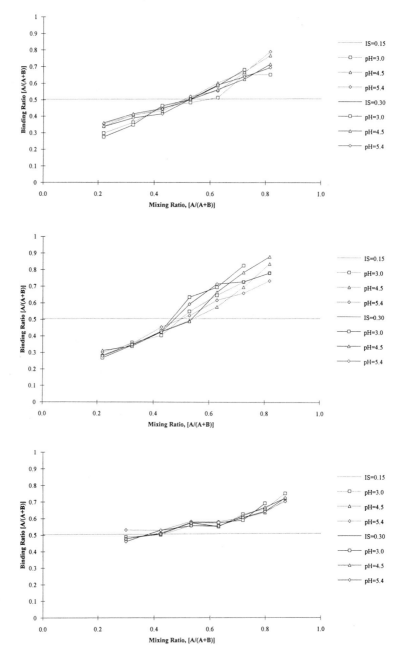

Figure 14.   Binding ratio in chitosan complexes as a function of chitosan and polyanion ratio in the reaction mixture. Adapted from (*115*). (a) Chitosan-alginate; (b) Chitosan-pectin; (c) Chitosan-carrageenan

6. Guilbert, S. In *Food Packaging and Preservation. Theory and Practice;* Mathlouthi, M., Ed.; Elsevier Applied Science Publishers Ltd.: London, England, **1986**; pp. 371.

7. Hatzidimitriu, E.; Guilbert, S. G.; Loukakis, G. *J. Food Sci.* **1987**, *52*, 472-474.

8. Torres, J. A. In *Water Activity: Theory and Applications to Food;* Rockland, L. B.; Beuchat, L. R., Eds.; Academic Press: New York, New York, **1987**; pp. 329.

9. Guilbert, S. G. In *L'emballage des denrèes alimentaires de grande consaommation;* Bureau, G.; Multon, J. J.; Biquet, B. Eds.; Tech. et. Doc. Lavoisier: Paris, France, **1989**, pp. 320-359.

10. Vojdani, F.; Torres, J. A. *J. Food Proc. Pres.* **1989**, *12*, 33.

11. Vojdani, F.; Torres, J. A. *J. Food Proc. Pres.* **1989**, *13*, 417.

12. Vojdani, F.; Torres, J. A. *J. Food Sci.* **1990**, *55*, 841-846.

13. Krotcha, J. M.; Hudson, J. S.; Camirand, W. M.; Pavlath, A. E. Paper No. 88-6523 presented at the ASAE Winter Meeting: Chicago, IL, **1988**.

14. Krotcha, J. M.; Hudson, J. S.; Avena-Bustillos, R. J. Presented at the IFT Annual Meeting, paper no. 762: Anaheim, CA, **1990**.

15. Krotcha, J. M.; Pavlath, A. E.; Goodman, N. In *Engineering and Food;* Spiess, W. E. L.; Schubert, H., Eds.; Proceedings of the Fifth International Congress on Engineering in Food; Elsevier Applied Science Publishing Co.: London, England, **1990**, Vol. 2.

16. Gennadios, A.; Weller, C. L. *Food Technol.* **1990**, *44*, 63-68.

17. Rico-Peña, D. C.; Torres, J. A. *J. Food Sci.* **1990**, *55*, 1468-1469.

18. Rico-Peña, D. C.; Torres, J. A. *J. Food Proc. Eng.* **1990**, *13*, 125-133.

19. Rico-Peña, D. C.; Torres, J. A. *J. Food Sci.* **1991**, *56*, 497-499.

20. Hagenmaier, R. D.; Shaw, P. E. *J. Agric. Food Chem.* **1990**, *38*, 1799-1803.

21. Hagenmaier, R. D.; Shaw, P. E. *J. Agric. Food Chem.* **1991**, *39*, 825-829.

22. Hagenmaier, R. D.; Shaw, P. E. *J. Agric. Food Chem.* **1991**, *39*, 1705-1708.

23. Avena-Bustillos, R. J. University of California, Davis, Ph.D. thesis, **1984**.

24. Martin-Polo, M.; Mauguin, C.; Voilley, A. *J. Agric. Food Chem.* **1992**, *40*, 407-412.

25. Martin-Polo, M.; Mauguin, C.; Voilley, A. *J. Agric. Food Chem.* **1992**, *40*, 413-418.

26. Mahmoud, R.; Savello, P. A. *J. Dairy Sci.* **1992**, *75*, 942-946.

27. Mahmoud, R.; Savello, P. A. *J. Dairy Sci.* **1993**, *76*, 29-35.

28. Gennadios, A.; Weller, C. L.; Testin, R. F. *J. Food Sci.* **1993**, *58*, 212-214.

29. Torres, J. A. Massachusetts Institute of Technology, Cambridge, Ph.D. thesis, **1984**.

30. McElhaney, R. N.; De Gier, J.; Van Deenen, L. L. M. *Biochim. Biophys. Acta* **1981**, *219*, 245.

31. Kamper, S. L.; Fennema, O. *J. Food Sci.* **1984**, *49*, 1478.

32. Kamper, S. L.; Fennema, O. *J. Food Sci.* **1984**, *49*, 1482.

33. Kamper, S. L.; Fennema, O. *J. Food Sci.* **1985**, *50*, 382.

34. Brake, N. C.; Fennema, O. R. *J. Food Sci.* **1993**, *58*, 1422-1425.
35. Fennema, O. R.; Donhowe, T. G.; Fester, J. J. *Food Aust.* **1993**, *45*, 521-525.
36. Nelson, K. L.; Fennema, O. R. *J. Food Sci. Off. Publ. Inst. Food Technol.* **1991**, *56*, 504-509.
37. Torres, J. A. In *Protein Functionality in Food Systems;* Hettiarachchy, N. S.; Ziegler, G. R., Eds.; Marcel Dekker, Inc.: New York, New York, **1994**; pp. 467-507.
38. *Edible Coatings and Films to Improve Food Quality;* Krochta, J. M.; Baldwin, E. A.; Nisperos-Carriedo, M. Eds.; Technomic Pub. Co.: Lancaster, PA, **1994**.
39. Ingram, M.; Dainty, R. H. *J. Appl. Bacteriol.* **1971**, *34*, 21.
40. Vitkov, M. *Veteranarnomeditsinski Nauki* **1973**, *10*, 55.
41. Vitkov, M. *Veteranarnomeditsinski Nauki* **1974**, *11*, 17.
42. Dainty, R. H.; Shaw, B. G.; De Boer, K. A.; Scheps, E. S. J. *J. Appl. Bacteriol.* **1975**, *39*, 73.
43. Gill, C. O. *J. Appl. Bact.* **1979**, *47*, 367.
44. Anderson, M. E.; Sebaugh, J. L.; Marshall, R. T.; Stringer, W. C. *J. Food Protect.* **1980**, *43*, 21.
45. Leistner, L.; Roedel, W.; Krispien, K. In *Water Activity: Influences on Food Quality;* Rockland, L. B.; Stewart, G. F., Eds.; Academic Press: New York, NY, **1981**; pp. 855.
46. Maxcy, R. B. *J. Food Protect.* **1981**, *44*, 550.
47. Anon *Marine Fisheries Rev.* **1983**, *45*, 24.
48. Cunningham, F. E. *J. Food Sci.* **1976**, *44*, 863.
49. Robach, M. C. *J. Food Protect.* **1979**, *42*, 855.
50. Bone, D. P. In *Water Activity: Theory and Applications to Food;* Rockland, L. B.; Beuchat, L. R., Eds.; Academic Press: New York, NY, **1987**; pp. 369.
51. Grundke, G.; Kuklov, K. *Lebensmittelindustrie* **1980**, *27*, 13.
52. Torres, J. A.; Bouzas, J. O.; Karel, M. *J. Food Proc. Pres.* **1985**, *9*, 93.
53. Stiles, M. E.; Ng, L. K. *J. Food Protect.* **1979**, *42*, 464.
54. Scott, W. J. *Austr. J. Biol. Sci.* **1953**, *6*, 549.
55. Robach, M. C.; Ivey, F. J. *J. Food Protect.* **1978**, *41*, 284.
56. D'Aubert, S.; Politi, P. G.; Simonetti, P. *Industrie Alimentari* **1980**, *19*, 759.
57. Holley, R. A. *Appl. Environ. Microbial.* **1981**, *41*, 422.
58. Robach, M. C.; Sofos, J. N. *J. Food Protect.* **1982**, *45*, 374.
59. Lueck, E. *Fleischwirtsch* **1984**, *64*, 727.
60. Greer, G. C. *J. Food Protect.* **1981**, *45*, 82.
61. Crank, J. In *The Mathematics of Diffusion;* 2nd ed.; Clarendon Press: Oxford, England, **1976**.
62. Rogers, C. E. In *Polymer Permeability;* Comyn, J., Ed.; Elsevier Applied Science Publishers Ltd.: London, England, **1985**; pp. 11.
63. Karel, M. In *Physical Principles of Food Preservation;* Karrel, M.; Fennema, O. R.; Lund, D. B., Eds.; Principles of Food Science. Part II; Marcel Dekker, Inc.: New York, NY, **1975**; pp. 399.

64. Colton, C. K.; Smith, K. A.; Merrill, E. W.; Farrell, P. C. *J. Biomed. Mater. Res.* **1971**, *5*, 459.

65. *Theory, Determination and Control of Physical Properties of Food Materials;* Rha, C., Ed.; D. Reidel Publishing Co.: Dordrecht, Holland, **1975**.

66. *Perry's Chemical Engineers' Handbook*; Perry, R. H.; Green, D., Eds.; 6th ed.; McGraw Hill, Inc.: New York, NY, **1984**.

67. Newman, A. A. *Glycerol;* C.R.C. Press: Cleveland, OH, **1968**.

68. De Gier, J.; Mandersloot, J. G.; Van Deenen, L. L. M. *Biochim. Biophs. Acta* **1968**, *150*, 666.

69. Hiemenz, P. C. In *Principles of Colloid and Surface Chemistry;* 1st ed.; Marcel Dekker, Inc.: New York, NY, **1977**; pp. 125.

70. Donnan, F. G. *Zeitschrift fur Elektrochemi* **1911**, *17*, 572.

71. Donnan, F. G. *Chemical Reviews* **1924**, *1*, 73.

72. Donnan, F. G.; Guggenheim, E. A. *Z. Physikal Chem. Abt. A.* **1932**, *162*, 346.

73. Donnan, F. G. *Z. Physikal Chem. Abt. A.* **1934**, *168*, 24.

74. Scallan, A. M.; Grignon, J. *Svensk papperstidning* **1979**, *2*, 40.

75. Grignon, J.; Scallan, A. M. *J. Appl. Polymer Sci.* **1980**, *25*, 2829.

76. Bade, M. L.; Wick, R. L. In *Biologically Active Natural Products. Potential Use in Agriculture*; Cutler, H. G., Ed.; ACS Symp. Series 380; Am. Chem. Soc.: Washington, DC, **1988**.

77. Allan, C.; Hadwiger, L. A. *Environ. Mycol.* **1979**, *3*, 285-287.

78. Stossel, P.; Leuba, J. L. *Phytopath. Z.* **1984**, *111*, 82-90.

79. Hirano, S.; Nagao, N. *Agric. Biol. Chem.* **1989**, *53*, 3065-3066.

80. Evans, E. E.; Kent, S. P. *J. Histochem. Cytochem.* **1962**, *10*, 24-28.

81. Leuba, S.; Stossel, P. In *Chitin in Nature and Technology;* Muzzarelli, R. A. A.; Jeniaux, C.; Gooday, C., Eds.; Plenum Press: New York, NY, **1985**; pp. 217.

82. Young, D. H.; Kohle, H.; Kauss, H. *Plant Physiol.* **1982**, *70*, 1449-1454.

83. Young, D. H.; Kauss, H. *Plant Physiol.* **1983**, *73*, 698.

84. Siegel, S. M.; Daly, O. *Plant Physiol.* **1972**, *41*, 1429-1434.

85. Magnolta, D. *U.K. Patent Application GB 2029688A.* **1978**.

86. Imeri, A.G.; Knorr, D. *J. Food Sci.* **1988**, *53*,1707.

87. Savant, V.D.; Torres, J. A. *American ChitoScience Society* Newsletter, December **1995**, *1*, 1.

88. Austin, P. R.; Brine, C. J.; Castle, J. E. Zikakis, J. P. *Science* **1981**, *212*, 749.

89. Landes, D. R.; Bough, W. A. *Bull. Environ. Contam. Toxicol.* **1976**, *15*, 555.

90. Zikakis, J. P.; Saylor, W. W.; Austin, P. R. In *Chitin and Chitosan;* Hirano, S.; Tokura, S., Eds.; Japanese Soc. Chitin and Chitosan, Sapporo, Japan, **1982**.

91. Knorr, D. *Process Biochem.* **1986**, *6*, 90.

92. Vorlop, K. D.; Klein, J. *Biotechnol. Lett.* **1981**, *3*, 9.

93. Klein, J.; Vorlop, K. D. In *Foundation of Biochemical Engineering: Kinetics and Thermodynamics in Biological Systems*; Blanch, H. W.; Papoutsakis, E. T.; Stephanopoulos, G., Eds.; American Chemical Society: Washington, DC, **1983**.

94. Knorr, D.; Teutonico, R. A. *J. Agric. Food Chem.* **1986**, *34*, 96.

95. Knorr, D. In *Biotechnology of Marine Polysaccharides;* Colwell, R. R.; Parisier, E. R.; Sinskey, A. J., Eds.; Hemisphere Publishing Co.: Washington, DC, **1985**.

96. Stanley, W. L.; Watters, G. G.; Chan, B.; Mercer, J. M. *Biotechnol. Bioeng.* **1975**, *17*, 315.

97. Muzzarelli, R. A. A. *Enzyme Microb. Technol.* **1980**, *2*, 177.

98. Muzzarelli, R. A. A.; Parisier, E. R. *Proceedings of 1ˢᵗ International Conference on Chitin and Chitosan.* MIT Sea Grant Program, Massachusetts Inst. of Technology: Cambridge, MA, **1978**.

99. Senstad, C.; Mattiasson, B. *Biotechnol. Bioeng.* **1989**, *33*, 216.

100. Yoshida, H.; Nishihara, H.; Kataoka, T. *Biotech. & Bioeng.* **1994**, *43*, 1087.

101. *Chitin*; Muzzarelli, R. A. A., Ed.; Pergamon Press: Oxford, **1977**; pp. 321.

102. Fisher, D. *Food Business*, **1989**, May 22 issue.

103. McKay, G.; Blair, H. S.; Gardner, J. R. *J. Appl. Polym. Sci.* **1982**, *27*, 3043.

104. Volesky, B. *Trends Biotechnol.* **1987**, *5*, 96.

105. Chavasit, V.; Torres, J. A. *Biotechnol. Progress.* **1990**, *6*, 2.

106. Soto-Peralta, N. V.; Muller, H.; Knorr, D. *J. Food Sci.* **1989**, *54*, 495.

107. Bough, W. A.; Landes, D. *J. Dairy Sci.* **1976**, *59*, 1874.

108. Bough, W. A. *J. Food Sci.* **1975**, *40*, 297.

109. Knorr, D. *Food Technol.* **1991**, *45*, 114.

110. Castillo, F. J. In *Yeast Biotechnology and Biocatalysis*; Verachtert, H.; De Mot, R., Eds.; Marcel & Dekker: New York, NY, **1990**; pp. 297.

111. Gonzalez Siso, M. I. *Bioresource Technol.* **1996**, *57*, 1.

112. Gardner, D. *Modern Dairy* **1989**, *68*, 15.

113. Kosikowski, F. V. *J. Dairy Sci.* **1979**, *62*, 1149.

114. Selmer-Olsen, E.; Ratnaweera, H. C.; Pehrson, R. *Water Sci. Technol.* **1996**, *34*, 33.

115. Mireles, C.; Martino, M.; Bouzas, J.; Torres, J. A. In *Advances in Chitin & Chitosan*; Brine, C. J.; Sanford, P. A.; Zikakis, J. P., Eds.; Elsevier Applied Sci.: New York, NY, **1992**; pp. 506.

116. Chavasit, V.; Torres, J. A. *Biotech. Progress.* **1990**, *6(1)*, 2.

117. Savant, V. D.; Torres, J. A. *Annual Meeting of the Inst. of Food Technol.*; Atlanta, GA, **1998**.

Chapter 18

Effect of Biopolymers on the Formation, Drying Kinetics, and Flavor Retention of Tamarind (*Tamarindus indica*) Foam-Mats

H. Romero-Tehuitzil[1], E. J. Vernon-Carter[1], M. G. Vizcarra-Mendoza[1], and C. I. Beristaín[2]

[1]Departamento Ingeniería de Procesos e Hidraúlica, Universidad Autónoma Metropolitana-Iztapalapa, A.P. 55-534, D.F. Mexico
[2]Instituto de Ciencias Básicas, Universidad Autónoma Veracruzana, A.P. 575 Xalapa, Veracruz, Mexico

The quality of dehydrated fruit foam-mats is highly dependent on the characteristics and composition of the foaming agents employed, as these influence the drying kinetics and sensory flavor perception of the dried pulps. This work evaluated the use of three types of foaming agents (ovalbumin, a protein obtained from egg; mesquite gum, a polysaccharide; and a mixture of synthetic, low molecular weight, surface active compounds) on the drying kinetics of tamarind foam-mats and on the sensory flavor quality of the dry tamarind powder.

About 14 million tons of tropical fruits are produced annually worldwide (*1*), of which less than 17% is utilized industrially in food and non-food applications. Mexico is considered a major producer of tropical and subtropical fruits in the world (*2*). Also, Mexico is the second largest consumer of soft drinks (*3*) and around 4-7% of tropical fruit products are incorporated as fruit pulps in soft drinks and other beverages.

Traditionally, fruit pulp is kept in cold storage after addition of the prerequisite amounts of chemical preservatives. This results in very high energy costs and in variable quality characteristics, due to fruit stationality and to pulp rotation in cold storage.

Spray-drying frequently has been used for preserving fruit pulps. However, the method requires the use of high drying temperatures and very large amounts of carrier materials, necessary for imparting an adequate solids content and fluidity to the pulps being processed. In consequence, a dry product with diminished sensory characteristics is obtained (*4*).

Foam-mat drying of fruit pulps could become a promising drying method if two main problems are overcome. First, low drying surface area makes the process economically infeasible. In this regard, Romero-Tehuitzil *et al.* (*5*) proposed a fluidized bed system with an extended drying area provided by inert suspended spheres covered by the fruit foam. Second, the use of synthetic, low molecular weight, surface active compounds as foaming agents produces an undesirable aftertaste in the dried products at concentrations as low as 0.5% (w/w). It may be possible to substitute other surface active materials, such as biopolymers, to achieve improved aftertaste in dried products for this purpose.

With this in mind, the main objectives of this work were two-fold: 1) to establish if the protein, ovalbumin, and the polysaccharide, mesquite gum, could substitute for the use of synthetic, low molecular weight, surface active compounds as foaming agents of tamarind pulp; and 2) to establish if the foaming agents effect the drying kinetics of tamarind fruit pulp and sensory quality of dried tamarind powder.

Experimental Methods

Materials. Tamarind pulp (*Tamarindus indica*), supplied by Sociedad Cooperativa Trabajadores de Pascual, S.C.L. (México), was passed through a 0.5 mm sieve. Particles having a larger size were discarded. The employed tamarind pulp was standardized to a soluble solids content of 16°Brix with distilled water as determined by a Bausch and Lomb ABBE-3L refractometer.

Three foaming agents of different nature and composition were employed. First, was a mixture of synthetic, low molecular weight, surface active compounds (SAC) consisting of Span 60 (sorbitan monostearate) and Tween 80 (polyoxyalkylene derivate of sorbitan monooleate). It was purchased from Canamex, S.A. de C.V. (México), and has a Hydrophilic-Lipophilic Balance (HLB) of 8.0, calculated by the procedure of Dziezack (6). Second, was egg protein ovalbumin (OA), provided by Alimentos Deshidratados "Campeón", S.A. de C.V. (México). It is a globular phosphoglycoprotein in the form of a dry powder with the following specifications: 80-87% protein, Nx6.79, contains 4 SH groups and 2 disulfide bond, molecular weight of 45,000 daltons, isoelectric point at pH 4.6, phosphorus content of 0-1.3% and is fat-free. Ovalbumin is sensitive to surface denaturation, which makes it a good foam stabilizer (7). Third, was mesquite gum (MG), the exudate of a leguminous tree (*Prosopis juliflora*). It is a highly branched heteropolymolecular complex polysaccharide, bearing a protein content whose concentration varies with botanical origin (1.2-5.8%). The primary structure of the carbohydrate component of mesquite gum has been described as a core of β-D-galactose residues, comprising a $(1\rightarrow3)$-linked backbone with $(1\rightarrow6)$-linked branches. Side chains contain L-arabinose, L-rhamnose, β-D-glucuronate and 4-O-Me-β-D-glucuronate (8).

Experimental Design. The foaming agents (SAC, OA and MG) were applied to the tamarind pulp according to a 2^k factorial experimental design, where k = 3 and represents the three mentioned foaming agents at a maximum and minimum level (9). The maximum level of the foaming agents was selected in accordance to reported usages (7,10,11). Thus, the value for ovoalbumin was 1.0% (w/w), for mesquite gum 3.0% (w/w) and surface active compounds 0.5% (w/w). The minimum level is represented by the absence of the foaming agent.

Foam Formation. The foaming agents, previously hydrated in water to 50% (w/w) solutions, were added to 100 g tamarind pulp samples in 600 ml beakers, according to the factorial experimental design. The different formulations were then whipped with an Oster mixer model 4450-8 (Sunbeam Continental, Ltd.) at 4600 rpm, maintained at 25°C with the aid of a regulated water bath (Brookfield TC-500), until foams with constant density were obtained (*12*). The foams were evaluated for mean drainage time, density and percentage volume increase (overrun).

Foam Drainage. Foam drainage mean time was determined using a 10 ml pipette which was filled with the tamarind foam, sealing the top and bottom of the pipette. The drained liquid (ml) was measured as a function of time. The foam stability was associated with the necessary time to drain half of the foam volume *(13)*.

Foam Density. Foam density was measured gravimetrically as described by Beristaín *et al. (10)*.

Foam Overrun. Foam percentage volume increase or overrun was measured *in situ* by the procedure developed by Cheftel *et al. (14)*. Foam overrun is a way of determining the foaming capacity of a foaming agent.

Foam-Mat Drying. All the tamarind fruit foams batches, even the unstable ones, were dried at 50 and 60°C in an Ohaus moisture balance model MB200 with a 0.001 g sensitivity. 10 g of tamarind foam was put in aluminum trays (11.5 cm diameter) with a foam-mat thickness of 1.5-2.0 mm. Moisture loss was directly read from the moisture balance digital display every 5 min for 2 hr.

Drying Kinetics. Drying kinetics were determined by plotting the change in moisture content of the tamarind foam-mat (g of water/g dry solids) with drying time. The constant drying period was determined from the linear part of the curves, applying linear regression to the experimental points and rejecting those that did not comply with an $R = 0.995$. The last point in the linear regression was considered as the moisture critical point or the beginning of the falling rate drying period.

Foam and Dry Powder Micrographs. Micrographs (100X) of selected tamarind foams and dry powder were taken with a Zeiss compound optical microscope model 476005-9901 (West Germany) coupled to a Contax camera model 139 Quartz (Japan).

Sensory Evaluation. In order to establish if the different treatments of dried tamarind foam presented differences in their flavor retention, a sensory analysis was applied. It consisted of a verbal hedonic scale of 5 points, which included the terms: I like it very much, I like it moderately, I don't like it or dislike it, I dislike it moderately and I dislike it extremely. All panelists were staff or students of the Biotechnology Department of the Universidad Autónoma Metropolitana-Iztapalapa, who had previous sensory analysis experience. The different tamarind foam treatments were dried in a tray drier at 50°C. The resulting tamarind powders were rehydrated (0.5 g in 25 ml water) and given to the panelists for evaluation at room temperature.

Statistical Analysis. The foam mean drainage time, overrun and density results were subjected to an analysis of variance (ANOVA), using Duncan's multiple range test, to determine whether differences among means were statistically significant at the 5% level (*15*). The sensory evaluation results were subjected to an ANOVA, followed by Tukey's test, to determine if a least significant difference among the treatments existed at the 1% level (*16*).

Results and Discussion

Foam Evaluation. The drainage mean time, density and overrun of the different foam treatments are shown in Table I.

Table I. Tamarind Foam Mean Drainage Time, Density and Overrun as a Function of Foaming Agents Composition, at 25°C

Foaming Agents	Mean Draining Time (hr)	Density (g/cm³)	Overrun (%)
OA	3.12[e]	0.847[e]	50[e]
MG	0.137[f]	0.924[f]	30[f]
SAC	6.57[d]	0.665[d]	90[d]
OA-MG	7.93[c]	0.552[c]	101[c]
OA-SAC	10.03[b]	0.492[b]	123[b]
MG-SAC	12.33[a]	0.375[a]	180[a]
OA-MG-SAC	9.83[b]	0.478[b]	133[b]

[a,b,c,d,e,f]Figures in a common column with varying superscripts are significantly different (p < 0.05). OA = ovalbumin, MG = mesquite gum, SAC = blend of Tween 80 - Span 60.

Figure 1 shows a selection of micrographs of different foam treatments before and after drying. Figure 1a depicts the foam made with treatment MG-SAC, which exhibited the highest stability. It was characterized by small, homogeneous and spherical bubbles, indicative of a strong, elastic film capable of resisting the pressure

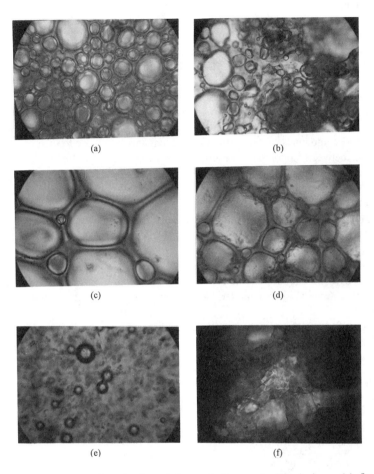

Figure 1. Micrographs of selected foams before and after drying. (a) foam treatment MG-SAC, (b) MG-SAC dry powder, (c) foam treatment OA-MG, (d) OA-MG dry powder, (e) foam treatment MG, and (f) MG dry powder.

of the enclosed air (*17*). Figure 1c shows the foam made with the OA-MG blend. This foam had a slightly better stability than that made with SAC, but substantially less stability than that made with MG-SAC. These results were reflected by the foam morphology, which was characterized by large polyhedrical bubbles and a thin interfacial film. This thinning-out was accompanied by a depletion of foaming agents and, thus, by a "sucking-in" phenomenon of the bubbles, which, in turn, caused instability and draining-out of the interlamellar liquid (*17*). Finally, Figure 1d shows the morphology of the MG formulation, the most unstable of the foams. It was characterized by a polydisperse and heterogeneous size distribution of bubbles. This gave rise to different internal bubble pressures and to an intrinsically unstable foam.

Thus, the above results point out that biopolymers and low molecular weight, surface active compounds tend to form more stable foams than those formed by binary interactions between biopolymers, or tertiary interactions.

The mean drainage time, overrun and density of the different treatments were significantly different ($p < 0.05$), with the exception of treatments OA-SAC and OA-MG-SAC (Table I).

The stability of the different foam treatments from lower to higher was: MG < OA < SAC < OA-MG < OA-GM-SAC, OA-SAC < MG-SAC.

Drying Kinetics. All the different foam-mats, even those considered as unstable (treatments MG and OA), were dried at 50 and 60°C. However, as the drying kinetics at both temperatures was rather similar, the following discussion is based on the results obtained at 50°C.

Figure 2 shows the water loss versus drying time plots for the different foam-mat treatments. The curves all exhibited different drying kinetics, mainly in regards to the duration of the constant drying period and the onset of the decreasing drying period.

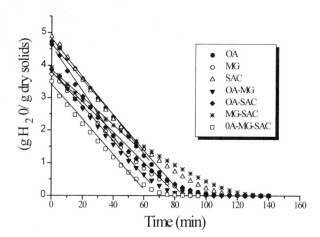

Figure 2. Drying kinetics of different tamarind foam-mat treatments at 50°C. Solid lines (—) depict the duration of the constant drying period for each treatment.

**Table II. Onset of the Drying Critical Time for the Different
Foam-Mat Treatments, at 50°C**

Foam-Mat Foaming Agency Composition	Onset of Critical Time (min)
OA	90
MG	80
SAC	65
OA-MG	65
OA-SAC	45
MG-SAC	55
OA-MG-SAC	60

Table II shows the time at which the end of the constant drying period (critical time) was reached. Foam-mats derived from the two biopolymer (OA and MG) had longer constant drying periods and shorter critical times than the rest of the composition. OA and MG formulations were the most unstable, forming poor foams which tended to drain the interlamellar water very shortly after foam formation.

Thus, most of the drying rate took place in the constant drying period, as the water evaporated as practically "free water." The foam-mat obtained from the mixture of Tween 80-Span 60 had a considerably shorter constant drying period than the OA and MG treatments, indicating that the diffusion of interlamellar trapped water through the bubble film structure was hindered. This, thus, forced the onset of the critical time to take place at shorter drying times.

The binary and the tertiary formulations, all show shorter critical times than the SAC treatment, which indicates that these interactions give rise to more structured films than single components do. These synergistic interactions of different components, besides providing stability against foam collapse, provide a barrier that retards the diffusion of water to the evaporative surface and retain volatile compounds responsible for flavor within the interlamellar structure. Bangs and Reineccius (18) and Yamamoto and Sano (19) found that the shorter the onset of the critical time, the better the sensory characteristics of the dried product because it better retains the volatile compounds responsible for taste and aroma. It is known that interactions between biopolymers and low molecular weight surface active compounds promote the formation of minimum critical thickness films of high elasticity, which enhance foam stability, and of reticulated structures with varying degrees of tortuosity, which hinder the diffusion of water and volatile compounds to the evaporative surface (20). From the obtained results, it appears that biopolymer-low molecular weight surface active compound (MG-SAC and OA-SAC) interactions provide more reticulated and tortuous structures than do biopolymer-biopolymer interactions (MG-OA and MG-OA-SAC).

Figure 1 also shows the micrographs of the dry tamarind powders obtained from a stable foam (b), a fairly stable foam (d) and an unstable foam (e). Note, the morphology of the three powders is completely different, which should, in consequence, effect physicochemical behavior. This aspect is beyond the scope of this study, however.

Sensory Evaluation. Statistical analysis indicates the judges found that the different rehydrated tamarind foam-mats were significantly different among themselves at $p < 0.01$. Furthermore, by estimating the least significant difference (L.S.D.) among the treatments, it was possible to order the judges flavor perception in decreasing order as follows: MG-SAC (0.407) > OA-SAC (0.367) > OA-MG(0.167) > OA(0.031) > MG(0.011) > SAC(-0.340) > OA-MG-SAC(-0.940). The parenthetic values indicate the relative flavor perception of the different samples, where a positive value indicates "I like it," a zero value indicates "I neither like it or dislike it" and a negative value indicates "I dislike it."

These results indicate that the dried foam-mats obtained from treatments MG-SAC, OA-SAC and OA-MG have an acceptable flavor, while the OA and MG treatments were perceived as not having a flavor (plain flavor) and, finally, treatments SAC and OA-MG-SAC have an unacceptable flavor. These results also confirm that the low molecular weight, surface active compounds provide an unacceptable flavor to the dry foam-mats. However, by combining them with biopolymers, this limitation can be overcome. Furthermore, unstable foams tend to produce insipid, plain flavored powders, as all of the flavoring agents are drained-out.

Conclusions

Foam stability can be controlled by selecting the nature and composition of the foaming agents employed. Interactions between biopolymers and low molecular weight surface active compounds have a synergistic effect which results in a longer mean drainage time, increased overrun and lower density tamarind fruit foams. Furthermore, the drying kinetics of the fruit foam may be controlled by designing "taylor made" interfacial bubble films that hinder the diffusion of interlamellar held water and volatile compounds to the evaporative surface, thus obtaining a dry product with enhanced sensory flavor attributes.

Acknowledgment

The authors wish to thank the Consejo Nacional de Ciencia y Tecnologia (CONACyT) of México, for partially financing this project through agreement 25153-B.

Literature Cited

1. INEGI *Anuario Estadístico de los Estados Unidos Mexicanos*; México, **1995**; pp. 262.

2. Macro Asesoría Económica *Análisis de Riesgo de 80 Industrias Durante 1997-1998 en México*; México, **1997**; pp. 13-15, 37-39.
3. Euromonitor *Consumer Mexico*: London, **1996**; pp. 92-96.
4. Beristaín, C. I.; Romero-Tehuitzil, H.; Vernon-Carter, E. J.; Vizcarra-Mendoza, M. G.; Ruíz-Martínez, R.S. *Advances Ing. Quím.* **1994**, *4*, 20-24.
5. Romero-Tehuitzil, H.; Vernon-Carter, E. J.; Beristain, C. I.; Vizcarra-Mendoza, M. G.; Ruíz-Martínez, R. S. In *Abstracts of the 9th World Congress of Food Science and Technology*: Budapest, Hungary, **1995**; Vol. 2, pp. 37.
6. Dziezak, J. D. *Food Technol.* **1988**, *42*, 172-186.
7. Baniel, A.; Fains, A.; Popineau, Y. *J. Food Sci.* **1997**, 377-381.
8. Goycoolea, F. M.; Calderón de la Barca, A. M.; Balderrama, J. R.; Valenzuela, J. R. *Intern. J. Biol. Macromol.* **1997**, *21*, 29-36.
9. Hasenhuettl, G. L. In *Food Emulsion and Foams: Theory and Practice;* Gaden, E. L., Ed.; AIChE Symposium Series 277; American Institute of Chemical Engineers: New, York, NY, **1990**; pp. 35-43.
10. Beristaín, C. I.; García, H.; Vázquez, A. *Drying Technol.* **1993**, *11*, 89-95.
11. Glicksman, M. In *Food Hydrocolloids;* Glicksman, M., Ed.; CRC Press, Inc.: Boca Raton, FL, **1983**, Vol. 2; pp. 7-29.
12. Romero-Tehuitzil, H. *Secado de Pulpa de Guayaba (Psidium guajava) por Fluidización*; M. Sc. Thesis in Chemical Engineering, Universidad Autónoma Metropolitana-Iztapalapa: México, **1995**.
13. Philips, L. G.; German, J. B.; O'neill, T. E.; Foegeding, E. A.; Harwalkar, V. R.; Kilara, A.; Lewis, B. A.; Mangino, M. E.; Morr, C. V.; Regestein, J. M.; Smith, D. M.; Kinsella, J. E. *J. Food Sci.* **1990**, *55*, 1441-1445.
14. Cheftel, J. C.; Cuq, J. L.; Lorient, D. *Proteínas Alimentarias*; Editorial Acribia, S.A.: Zaragoza, Spain, **1987**; pp. 52-60.
15. O'Mahoney, M. *Sensory Evaluation of Foods. Statistical Methods and Procedures*; Marcel Dekker, Inc.: New York, NY and Basel, **1986**; pp. 164-166.
16. Anzaldúa-Morales, A. *La Evaluación Sensorial de los Alimentos en la Teoría y la Práctica*; Editorial Acribia, S.A.: Zaragoza, Spain, **1994**; pp. 84-87.
17. Aubert, J. H.; Kraynik, M. A.; Rand, P. B. *Investigación y Ciencia* **1986**, 38-47.
18. Bangs, W. E.; Reineccius, G. A. *J. Food Sci.* **1990**, 1356-1358.
19. Yamamoto, S.; Sano, Y. *Drying Technol.* **1995**, *13*, 29-41.
20. Krochta, J. M. In *Food Emulsion and Foams: Theory and Practice;* Gaden, E. L., Ed.; AIChE Symposium Series 277, American Institute of Chemical Engineers: New York, NY, **1990**; pp. 57-61.

INDEXES

Author Index

Subject Index

A